The Best Approximation Method in
Computational Mechanics

Theodore V. Hromadka II

The Best Approximation Method in Computational Mechanics

With 35 Figures

Springer-Verlag
London Berlin Heidelberg New York
Paris Tokyo Hong Kong
Barcelona Budapest

Theodore V. Hromadka II, PhD, PhD, PH, RCE
California State University, Fullerton and Computational Hydrology Institute,
Irvine, USA

ISBN-13: 978-1-4471-2022-3   e-ISBN-13: 978-1-4471-2020-9
DOI: 10.1007/978-1-4471-2020-9

British Library Cataloguing in Publication Data
Hromadka, T.V.
  Best Approximation Method in
  Computational Mechanics
  I. Title
  620.100285
ISBN-13: 978-1-4471-2022-3

Library of Congress Cataloging-in-Publication Data
Hromadka, Theodore V.
  The best approximation method in computational mechanics /
Theodore V. Hromadka II.
     p.   cm.
  Includes bibliographical references and index.
  ISBN-13: 978-1-4471-2022-3
  1. Functional analysis.   I. Title.
QA320.H76   1992   92-33754
515'.7--dc20   CIP

© Springer-Verlag London Limited 1993
Softcover reprint of the hardcover 1st edition 1993

Typesetting: Camera ready by author
69/3830-543210 Printed on acid-free paper

To Laura

Acknowledgements

The author pays acknowledgements to Dr. C.C. Yen of Williamson & Schmid, Tustin, California, who carefully reviewed the manuscript many times, and also helped prepared several of the computer applications. Acknowledgements are also paid to Ms. Phyllis Williams, who typed and modified the various versions of the manuscript. Thanks are given to Mr. Bill Burchard, who prepared the several figures. And finally, thanks are given to my wife, Laura, who supported me throughout this project.

Contents

CHAPTER 1
TOPICS IN FUNCTIONAL ANALYSIS

1.0. Introduction

With the overwhelming use of computers in engineering, science, and physics, the ability to approximately solve complex mathematical systems of equations is almost commonplace. And yet, despite the vast quantities of synthetic data one sometimes isn't quite sure whether these approximations are valid. A nagging question haunts the analyst as to whether the extrapolation of true data just achieved by the computer program really represents reality, or merely represents some impossible result that was generated by a collection of small but accumulative errors in both analysis and computation.

In order to investigate the validity of the computational results, a return to mathematical analysis of the computational scheme may be necessary. Consequently computer modelers, both experienced and novices, need to become familiar with the bodies of mathematical literature generally classified as functional analysis and the more recent numerical analysis. Many questions regarding the validity and competence of computer program algorithms can be answered with usage of theorems in functional analysis. Issues regarding algorithm convergence and stability oftentimes can be addressed in terms of concepts in functional analysis.

Fortunately many of the important concepts of functional and numerical analysis can be communicated in a readable setting without use of elaborate proofs and derivations. Oftentimes, the fine details accounted for in the detailed proof are not at issue in the underlying space of functions that the analyst is implicity using to develop the approximation.

A goal of this book is to present possibly the more important and useful functional analysis concepts that may serve the computer modeler in his/her search for "truth". The book may serve as an introduction to a functional analysis course, or may also serve as an introduction to mathematical analysis of computer modeling algorithms. In any event, the book may direct the attention of computer modelers to the already established principles and results assembled in functional analysis.

1.1. Set Theory

Because the computer program can only handle a finite **set** of approximations or calculations, a goal in modeling efforts is to optimize the accuracy of the approximation. Such a goal requires a measure of accuracy to be optimized, which in turn requires a well defined collection of rules and definition of terms so that there is no confusion regarding what was optimized and how it was optimized.

The starting point is the definition of a **set** as being a collection of **elements.** An element s belongs to a set S is written as $s \in S$. If s is not in S, then we write $s \notin S$.

Computer data are often stored in data banks composed of millions of data entries. Sometimes, these data can be represented by an algorithm, which can be used as a substitute for the data storage and also to eliminate the retrieval and data sorting process. For example, rather than store a table of trigonometric data corresponding to the function $\sin\theta$, $0 \le \theta \le \pi$, one can simple generate the tabled value on an as-need basis if one knew that the table actually was the $\sin\theta$ function. Thus, the set S of $\sin\theta$ data can be described by the characterization

$$S = \{x \in \mathbf{R}: x = \sin\theta, \ 0 \le \theta \le \pi \}$$

which means S is the set of all real numbers x such that x is a value of the function $\sin\theta$ for $0 \le \theta \le \pi$.

If S contains no elements, then S is the **empty** or **null** set, written as $S = \emptyset$. If every element of a set S is also found in a set T, then S is a **subset** of T, and is noted by $S \subset T$. Set S is a proper subset of T if there exists a point $x \in T$ such that $x \notin S$. From the definitions, $\emptyset \subset S$, even when S is the empty set. Two sets S and T are **equal** if $S \subset T$ and $T \subset S$. Equality of sets is noted by $S = T$. Let $S \subset U$. Then the **complement** of S with respect to U is the set S^c where

$$S^c = \{x \in U: x \notin S\} \tag{1.1.1}$$

In (1.1.1) it is understood that S^c is defined with respect to U. U is referred to as the **universal** set. When necessary, the notation of S^c with respect to U may be given by $S^c = U\text{-}S$, where the minus sign indicates removal of all points in U that are in S.

Example 1.1.1

Set theory finds frequent use in probability and statistics. In analyzing the possible outcomes of a die toss, the universal set, U, is finite in size, and can be described by the indexed notation $U = \{a_1, a_2, \cdots, a_n\}$ where a_k is the k-th element of U, and n = 6. A subset of U is $S = \{2,3,5\}$. And $S^c = \{1,4,6\}$.

The **union** of two sets S and T, both sets contained in the universal set U, is defined as

$$S \cup T = \{x \in U: x \in S \text{ or } x \in T\}$$

where "or" is analogous to the computer programming logic OR statement.

The **intersection** of S and T, both sets in U, is defined as

$$S \cap T = \{x \in U: x \in S \text{ and } x \in T\}$$

where "and" is analogous to the computer programming logic AND statement.

If $S \cap T = \emptyset$, then S and T are said to be **disjoint.**

Example 1.1.2

From Example 1.1.1, $S \cap U = S$; $S \cap S^c = \emptyset$.

Example 1.1.3

$S \subset (S \cup T)$ because for every point $x \in S$, $x \in S$ or $x \in T$ and hence $x \in (S \cup T)$.

Theorem 1.1. Let R,S,T be subsets of U. Then the following properties hold:

(i)   If $R \subset S$ and $S \subset T$, then $R \subset T$

(ii) $R \cap S = S \cap R$; (commutative property of $\cap$)

(iii) $R \cup S = S \cup R$; (commutative property of $\cup$)

(iv) $R \cup \emptyset = R$; $R \cap \emptyset = \emptyset$

(v) $R \cap R^c = \emptyset$

(vi) $(R \cup S) \cup T = R \cup (S \cup T)$; (associative property of $\cup$)

(vii) $(R \cap S) \cap T = R \cap (S \cap T)$; (associative property of $\cap$)

(viii) $R \cap (S \cup T) = (R \cap S) \cup (R \cap T)$; (distributive property)

(ix) $R \cup (S \cap T) = (R \cup S) \cap (R \cup T)$

(x) $U^c = \emptyset$

(xi) $(R \cup S)^c = R^c \cap S^c$; (De Morgan's property)

(xii) $S \cup \emptyset = S$; $S \cup U = U$; $S \cap U = S$; $S \cup S^c = U$

Thought Problem 1.1

Let A be the set of all sets. Is A a subset of A?

In set theory, an ordering of the elements contained in a set S may sometimes be of importance. For example the **ordered set** $(a_1, a_2, \cdots, a_n)$ is an n-tuple of ordered elements, a_i, by virtue of the index subscript. Ordering may be described by use of an **index set,** I, such that $\{S_i : i \in I\}$ is an **indexed family** of sets. The most common index set, and the one considered hereafter in this book, is the set I being the positive integers,

$$I = \{1, 2, \cdots\}$$

It follows that the notation

$$\cup_{i=1}^{n} S_i = S_1 \cup S_2 \cup \cdots \cup S_n = \{x \in U : x \in S_i \text{ for some } i = 1,2,\cdots,n\}$$

and

$$\cap_{i=1}^{n} S_i = S_1 \cap S_2 \cap \cdots \cap S_n = \{x \in U : x \in S_i \text{ for all } i = 1,2,\cdots,n\}$$

Theorem 1.2. The Rules of De Morgan

Let $S_1, S_2, \cdots S_n$ be subsets of U, then

(i) $(\cup_{i=1}^{n} S_i)^c = \cap_{i=1}^{n} S_i^c$

(ii) $(\cap_{i=1}^{n} S_i)^c = \cup_{i=1}^{n} S_i^c$

It is oftentimes convenient to describe a sequence of unions or intersections of sets by use of an index or pointer algorithm that specifies a particular subset J of I,

$$\underset{i \in J}{\cup} S_i = \{x \in U : x \in S_i \text{ for some } i \in J\}$$

and

$$\underset{i \in J}{\cap} S_i = \{x \in U : x \in S_i \text{ for all } i \in J\}$$

In Thought Problem 1.1.1 it was demonstrated that some control on A is needed in order to avoid absurdity or arguments of contradiction. Such a control is embodied in the concept of a sigma-algebra or σ-algebra. A σ-algebra is a family of sets that satisfy certain rules which guarantee a certain "proper behavior".

DEFINITION 1.1.1. A family of subsets A of the set A is called a σ-algebra if

(i) $\emptyset \in A$; $A \in A$

(ii) If $H \in A$, then $H^c \in A$

(iii) If $\{S_i, i \in I\}$ is a sequence of sets such that each $S_i \in A$, then
$$\bigcup_{i=1}^{\infty} S_i \in A.$$

Example 1.1.4

From the rules of De Morgan, given $\{S_i, i \in I\}$ is a sequence of sets such that $S_i \in A$, where A is a σ-algebra, then
$$\bigcap_{i=1}^{\infty} S_i \in A.$$

Example 1.1.5

Let A be the two sets $\emptyset$ and A. Then A is a σ-algebra.

1.2. Functions

In solving problems, one attempts to derive an output that is related in some way to a set of input data. That is, there is a function or mapping, f, that assigns to each element or vector of a set S (this element may be multi-dimensional, composed of several data values) a specific element or vector of the set T by
$$f: S \to T$$
Thus, if element $s \in S$, then
$$f: s \to f(s) \in T$$
The set of all input elements, S, is called the **domain** of f while the set of all output elements, T, is called the **range**. Should S and T be such that for s_1 and s_2 any two elements of S, then $f(s_1) \neq f(s_2)$, then f is **one-to-one**. And if f is one-to-one then there is an **inverse** function f^{-1} that reverses what f did to $s \in S$ by
$$f^{-1}(f(s)) = s, \text{ or } f^{-1}(t) = s \text{ where } f(s) = t$$
It is recalled that set T is the range of f on domain S, and hence since T is completely "used up" by the function f on S, f is also said to be **onto**.

The **composition** function $f^{-1}(f(s))$ can be noted by $f^{-1}f$ and is the **identity** function $I_f = f^{-1}f$.

Example 1.2.1.

For $S = \{x : x > 0\}$, $e^{\ln x} = x$. Thus $e^{\ln x} = f^{-1}f(x) = x$ on the restricted domain $x > 0$.

Example 1.2.2

Because $e^x > 0$, $\ln e^x = x$ is defined for all real x, and $\ln e^x = f^{-1}f(x) = x$ on $\mathbf{R}$.

Example 1.2.3

Let $S = \{(a_1, a_2, a_3): a_i \in \mathbf{R}, \ i = 1, 2, 3 \}$. Define f by f: $(a_1, a_2, a_3) \to a_1 + a_2 + a_3$. Then the domain of f is $\mathbf{R}^3$, and the range of f is $\mathbf{R}^1$. Note that an element $s \in S$ has three dimensions (or degrees of freedom) whereas element of the range of f, $t \in T$, has only one dimension.

Example 1.2.4

Let S be the set of continuous functions of the interval $[0, \pi]$, denoted by $C[0, \pi]$. The function T defined by

$$T(f) = \left(\int_0^{\pi} |f(x)|^p dx \right)^{\frac{1}{p}} \quad \text{for } f \in S$$

is a function with domain, $C[0, \pi]$, and range $\mathbf{R}^+$ (the positive real numbers). That is

$$T: \ C[0, \pi] \to \mathbf{R}^+$$

Example 1.2.5

Let S and T be subsets of $\mathbf{R}^3$ where $s = (s_1, s_2, s_3)$ and $t = (t_1, t_2, t_3)$ are respective elements. Suppose

$$s_1 = t_2 \sin\theta + t_3 \cos\theta$$

$$s_2 = t_2 \cos\theta - t_3 \sin\theta$$

$$s_3 = \pi t_3$$

Then, in matrix form $\mathbf{s} = \mathbf{A}\,\mathbf{t}$ where

$$\begin{pmatrix} s_1 \\ s_2 \\ s_3 \end{pmatrix} \begin{matrix} = \\ = \\ = \end{matrix} \begin{bmatrix} 0 & \sin\theta & \cos\theta \\ 0 & \cos\theta & -\sin\theta \\ \pi & 0 & 0 \end{bmatrix} \begin{pmatrix} t_1 \\ t_2 \\ t_3 \end{pmatrix}$$

Note that $f: \mathbf{R}^3 \quad \mathbf{R}^3$, and that the mapping $\mathbf{A}$ is a function of argument θ.

Example 1.2.6

An interval of real numbers has one of the forms: $[a,b]$, $[a,b)$, $(a,b]$, (a,b). The **characteristic** function χ_j finds use in numerical approximations, and is defined on an interval I_j by

$$\chi_j(x) = \begin{cases} 1, & x \in I_j \\ \\ 0, & \text{elsewhere} \end{cases}$$

Given a finite set of m disjoint intervals (or overlapping only at the endpoints), a **step** function $\psi(x)$ is defined on the I_j by

$$\psi(x) = \sum_{j=1}^{m} k_j \, \chi_j(x)$$

1.3. Matrices

The vast majority of computational models of physical phenomena utilize approximations that involve the solution of matrix systems. Oftentimes, the matrix systems are frequently regenerated to represent changing material or problem domain conditions, or transitional effects on external influences and boundary conditions.

An mxn matrix, $\mathbf{A}$, is an m-row by n-column array of numbers, as used in a DIMENSION (M,N) programming statement. The data entry of row i and column j of matrix $\mathbf{A}$ is noted by a_{ij}. A **row** vector is a 1xn matrix. A mxn matrix with $a_{ij} = 0$ for all i and j is a **zero** matrix. The **identity** matrix is an mxm (or **square**) matrix with

$$a_{ij} = \begin{cases} 1; & i = j, \ i = 1,2,\cdots,m \\ \\ 0; & i \neq j, \ i = 1,2,\cdots,m, \ j = 1,2,\cdots,m \end{cases}$$

The **main diagonal** of a square matrix is the data entries a_{ii}, $i = 1,2,\cdots m$. A square matrix with all zero entries below the main diagonal (a_{ij}, $i>j$) is in **upper triangular** form. If the matrix has all zero entries above the main diagonal (a_{ij}, $i<j$) the square matrix is in **lower triangular** form.

Example 1.3.1

The following matrices are examples of the (i) a zero matrix, (ii) an identity matrix, (iii) an upper triangular form, (iv) a lower triangular form. Also shown are the matrix dimensions.

(i) $\begin{bmatrix} 0 & 0 & 0 & 0 \\ 0 & 0 & 0 & 0 \end{bmatrix}_{2x4}$ (ii) $\begin{bmatrix} 1 & 0 & 0 \\ 0 & 1 & 0 \\ 0 & 0 & 1 \end{bmatrix}_{3x3}$ (iii) $\begin{bmatrix} 1 & 2 & 3 \\ 0 & 4 & 5 \\ 0 & 0 & 6 \end{bmatrix}_{3x3}$

(iv) $\begin{bmatrix} 1 & 0 & 0 \\ 2 & 3 & 0 \\ 4 & 5 & 6 \end{bmatrix}_{3x3}$

A mxm matrix **A** that satisfies $a_{ij} = 0$ whenever $i \neq j$, is called a **diagonal** matrix, and can be expressed in the notation $A = \text{diag}\,(a_{11}, a_{22}, \cdots, a_{mm})$. The **transpose** A^T of arbitrary dimension mxn matrix **A** is given by the algorithm

$$a^T_{ij} = a_{ji}; \; i = 1,2,\cdots,m, \; j = 1,2,\cdots,n$$

In modeling transport phenomena such as heat transport or groundwater flow, among others, **symmetric** matrices often result where $A^T = A$. A matrix is **skew-symmetric** if $A^T = -A$. The **trace** of a mxm matrix is

$$\sum_{i=1}^{m} a_{ii}$$

Example 1.3.2

The following matrices are examples of a (i) a diagonal matrix; (ii) a symmetric matrix; (iii) a skew-symmetric matrix

(i) $\begin{bmatrix} \pi & 0 & 0 \\ 0 & e & 0 \\ 0 & 0 & i \end{bmatrix}$; (ii) $\begin{bmatrix} 1 & 2 & 3 \\ 2 & 4 & 5 \\ 3 & 5 & 6 \end{bmatrix}$; (iii) $\begin{bmatrix} 0 & 1 & 2 \\ -1 & 0 & -3 \\ -2 & 3 & 0 \end{bmatrix}$

1.4. Solving Matrix Systems

Because many computer modeling techniques involve matrix systems of linear equations, finding solutions to these systems are an important problem. Consider the linear system

$$y_1 = a_{11}x_1 + a_{12}x_2 + \cdots + a_{1m}x_m$$
$$y_2 = a_{21}x_1 + a_{22}x_2 + \cdots + a_{2m}x_m \qquad (4.1)$$
$$\vdots \qquad\qquad \vdots$$
$$y_n = a_{n1}x_1 + a_{n2}x_2 + \cdots + a_{nm}x_m$$

which is written in matrix form as

$$y_{nx1} = A_{nxm}x_{mx1}$$

where

$$y_{nx1} = \begin{pmatrix} y_1 \\ y_2 \\ \cdot \\ \cdot \\ \cdot \\ y_n \end{pmatrix}_{nx1} ; \quad A_{nxm} = \begin{bmatrix} a_{11} & a_{12} \cdots & a_{1m} \\ a_{21} & a_{22} \cdots & a_{2m} \\ \cdot \\ \cdot \\ a_{n1} & a_{n2} \cdots & a_{nm} \end{bmatrix} ; \quad x_{mx1} = \begin{pmatrix} x_1 \\ x_2 \\ \cdot \\ \cdot \\ \cdot \\ x_m \end{pmatrix}$$

There are several methods available to find x, given y and A. Additionally, theorems prescribe whether solutions for x exist and, when solutions do exist, describe all the x vectors in the solution set. A cross-roads in the solution strategy is whether the matrix system is **homogeneous, $Ax = 0$; or non-homogeneous, $Ax = b \neq 0$.**

For a non-homogeneous matrix system, there are three possible solution scenarios:

Theorem 1.4.1

The system of linear equations (1.4.1) has three possible solution sets:

    (i)   no solution;

    (ii)   a unique solution vector;

    (iii)   an infinite number of solutions.

Example 1.4.1

Consider the system

$$\pi x_1 + e x_2 = 2$$
$$\pi x_1 + e x_2 = 2$$

or in matrix form

$$\begin{bmatrix} \pi & e \\ \pi & e \end{bmatrix} \begin{Bmatrix} x_1 \\ x_2 \end{Bmatrix} = \begin{Bmatrix} 2 \\ 2 \end{Bmatrix}$$

It is obvious that no new information is contained in the second equation, and that when $x_2 = s$, $x_1 = (2-es)/\pi$. The solution set is an infinite collection of vectors all of the form

$$\begin{Bmatrix} x_1 \\ x_2 \end{Bmatrix} = \begin{Bmatrix} (2-es)/\pi \\ s \end{Bmatrix} = \begin{Bmatrix} 2/\pi \\ 0 \end{Bmatrix} + s \begin{Bmatrix} -e/\pi \\ 1 \end{Bmatrix}$$

Note that if the row equations had been subtracted,

$$\begin{bmatrix} \pi & e \\ 0 & 0 \end{bmatrix} \begin{Bmatrix} x_1 \\ x_2 \end{Bmatrix} = \begin{Bmatrix} 2 \\ 0 \end{Bmatrix}$$

Example 1.4.2

Consider the system

$$\pi x_1 + e x_2 = 0$$
$$\pi x_1 + e x_2 = 1$$

or in matrix form

$$\begin{bmatrix} \pi & e \\ \pi & e \end{bmatrix} \begin{Bmatrix} x_1 \\ x_2 \end{Bmatrix} = \begin{Bmatrix} 0 \\ 1 \end{Bmatrix}$$

It is obvious that a contradiction exists between the two equations. Note that if we subtract row equations,

$$\begin{bmatrix} \pi & e \\ 0 & 0 \end{bmatrix} \begin{Bmatrix} x_1 \\ x_2 \end{Bmatrix} = \begin{Bmatrix} 0 \\ 1 \end{Bmatrix}$$

Example 1.4.3

Consider the system

$$x_1 + x_2 = 0$$
$$x_1 - x_2 = 1$$

or in matrix form

$$\begin{bmatrix} 1 & 1 \\ 1 & -1 \end{bmatrix} \begin{pmatrix} x_1 \\ x_2 \end{pmatrix} = \begin{pmatrix} 0 \\ 1 \end{pmatrix}$$

Subtracting rows,

$$\begin{bmatrix} 1 & 1 \\ 0 & -2 \end{bmatrix} \begin{pmatrix} x_1 \\ x_2 \end{pmatrix} = \begin{pmatrix} 0 \\ 1 \end{pmatrix}$$

giving $x_2 = -0.50$, $x_1 = +0.50$; or $(x_1, x_2)^T = 0.5 (1, -1)^T$

Several of the techniques used to determine the set of solution vectors to a matrix system (it is recalled that the set of solution vectors may be $\emptyset$) make use of **elementary row operations**. These operations are row manipulations that include:

(i)   interchange any two rows;

(ii) multiply a row by a nonzero scalar;

(iii) add the contents of a row (which can be multiplied by a nonzero scalar) to the contents of another row.

DEFINITION 1.4.1

Matrices A_{mxn} and B_{mxn} are **row equivalent** if B_{mxn} can be obtained from A_{mxn} by a finite number of elementary row operations.

DEFINITION 1.4.2

A matrix A_{mxn} is in **row echelon form** if

(i)   all zero rows follow the nonzero rows;

(ii) the first nonzero entry of each nonzero row is a 1 (a leading 1);

(iii) the leading one of each nonzero row is preceded by more zeros than the preceding nonzero row.

Example 1.4.4

Matrices (i) and (ii) are in row echelon form, and matrices (iii) and (iv) are not.

(i) $\begin{bmatrix} 1 & 0 & 5 \\ 0 & 1 & 3 \end{bmatrix}$ (ii) $\begin{bmatrix} 1 & 0 & 0 \\ 0 & 0 & 1 \\ 0 & 0 & 0 \\ 0 & 0 & 0 \end{bmatrix}$

(iii) $\begin{bmatrix} 0 & 0 & 0 \\ 0 & 1 & 0 \\ 0 & 0 & 0 \end{bmatrix}$
(iv) $\begin{bmatrix} 2 & 1 & 3 \\ 0 & 1 & 0 \\ 0 & 0 & 1 \end{bmatrix}$

The elementary row operations may be used to reduce matrix **A** into row echelon form while performing the same operations in tandem on the column vector, **b**. In this fashion, a solution set **x** may be obtained.

Example 1.4.5

Solve the matrix system

$$\begin{bmatrix} 1 & 2 & 3 \\ 4 & 4 & 2 \\ 2 & 1 & 2 \end{bmatrix} \begin{pmatrix} x_1 \\ x_2 \\ x_3 \end{pmatrix} = \begin{pmatrix} 1 \\ 0 \\ -1 \end{pmatrix} :$$

$$\begin{bmatrix} 1 & 2 & 3 \\ 0 & -4 & -10 \\ 0 & -3 & -4 \end{bmatrix} \begin{pmatrix} x_1 \\ x_2 \\ x_3 \end{pmatrix} = \begin{pmatrix} 1 \\ -4 \\ -3 \end{pmatrix}$$ add (-4) times first row to row 2; add (-2) times first row to row 3.

$$\begin{bmatrix} 1 & 2 & 3 \\ 0 & 1 & 5/2 \\ 0 & 1 & 4/3 \end{bmatrix} \begin{pmatrix} x_1 \\ x_2 \\ x_3 \end{pmatrix} = \begin{pmatrix} 1 \\ 1 \\ 1 \end{pmatrix}$$ divide row 2 by (-4); divide row 3 by (-3).

$$\begin{bmatrix} 1 & 2 & 3 \\ 0 & 1 & 5/2 \\ 0 & 0 & -7/6 \end{bmatrix} \begin{pmatrix} x_1 \\ x_2 \\ x_3 \end{pmatrix} = \begin{pmatrix} 1 \\ 1 \\ 0 \end{pmatrix}$$ add (-1) times row 2 to row 3.

$$\begin{bmatrix} 1 & 2 & 3 \\ 0 & 1 & 5/2 \\ 0 & 0 & 1 \end{bmatrix} \begin{pmatrix} x_1 \\ x_2 \\ x_3 \end{pmatrix} = \begin{pmatrix} 1 \\ 1 \\ 0 \end{pmatrix}$$ divide row 3 by (-7/6).

Then $x_3 = 0$, $x_2 = 1$, and $x_1 = -1$.

The solution set is the single vector $(x_1, x_2, x_3) = (-1, 1, 0)$ which can be verified by substitution into the original matrix system.

The Example 1.4.5 is analogous to Example 1.4.3 in that a unique solution was obtained. It is noted that in these two examples, the number of nonzero rows in the matrix A, in row echelon form, equals the number of unknowns. In comparison, Example 1.4.2 has no solution vector; and the number of nonzero rows of A, in row echelon form, is less than the number of nonzero rows in the b vector (counted downwards). Finally in Example 1.4.1, there are an infinity of solution vectors; and the number of nonzero rows of A, in row echelon form, is not exceeded by the number of nonzero rows in the b vector (counted downwards), and the number of unknowns exceed the number of nonzero rows of A in row echelon form. Because our key interest lies in solution of matrix systems which have a unique solution vector, a computer program can be prepared based upon using the elementary row operations. The **Gaussian elimination** method solves a matrix system by the above procedures, and is a widely used computational procedure in software products.

Because the elementary row operations are performed in tandem on both A and b, a useful technique is to augment the A matrix into A^* by adding the b vector as an additional column of A; A^* is called the **augmented** matrix. The **rank** of A is the number of nonzero rows of A in row echelon form. The rank of A^* is the number of nonzero rows of A^* in row echelon form.

Theorem 1.4.2

Let r = rank of A in row echelon form, and r^* = rank of A^* in row echelon form.

Let n = the dimension (number of rows) of vector x. The solution vectors of $Ax = b$ have three cases:

(i)   If $r < r^*$, there are no solution vectors;

(ii)   If $r = r^*$ and $r < n$, there are an infinity of solution vectors;

(iii)   If $r = r^*$ and $r = n$, there is a unique solution vector.

Note that in Theorem 1.4.2, necessarily $r \leq r^*$.

If a unique solution vector exists for $Ax = b$, then the inverse of A exists, A^{-1}. The solution vector is then determined by $x = A^{-1}b$. If A^{-1} exists, A is **nonsingular** and the determinant of A, $|A|$ or det A, is nonzero. Thus given m linear equations for m unknowns, if det $A_{m \times m}$ is nonzero, then a unique solution vector, x, exists.

14

Thought Problem 1.2

(1.) Consider the matrix system $A_{mxn} x_{nx1} = b_{mx1}$. The matrix A_{mxn} can be visualized as an assemblage of column vectors

$$A_{mxn} = \left[v_1, v_2, v_3, \cdots v_n \right]$$

where each v_k is a mx1 vector, $v_k = (a_{1,k}, a_{2,k}, \cdots, a_{m,k})^T$. Then the matrix system can be written as

$$x_1 v_1 + x_2 v_2 + \cdots + x_m v_n = b$$

where the values $x_1, x_2, \cdots, x_n$ are the data row entries of column vector x. What is the best choice of values for the x_i; i=1,2,$\cdots$,n such that the error of approximating vector b, with the set of vectors $\{v_i;$ i=1,2,$\cdots$,n$\}$ is minimum? How is the approximation error defined? Does the "best" choice of $\{x_i;$ i=1,2,$\cdots$,n$\}$ depend upon the definition of approximation error?

(2.) For a given selection of values of $\{\hat{x}_i,$ i=1,2,$\cdots$n$\}$, where hat notation will be used to designate approximate values of a variable. Let

$$\hat{b} = \sum_{i=1}^{n} \hat{x}_i v_i$$

Define approximation error, E_p, by

$$E_p = \left(\sum_{j=1}^{m} \left| \hat{x}_i - b_j \right|^p \right)^{\frac{1}{p}}, \; p > 0$$

where $\hat{b}_j$ and b_j are row j data entries (of m rows) of $\hat{b}$ and b, respectively. Does the optimum choice of the $\{x_i\}$ depend upon the choice for p?

(3.) Suppose that $Ax = b$ has a unique solution vector. Re-answer part (2) given this additional fact.

The matrix system $\mathbf{A}\mathbf{x} = \mathbf{b}$ is homogeneous if $\mathbf{b}$ is the zero vector $\mathbf{0}$. In this case, there are only two possible solution sets of vectors; namely, the **trivial** solution or zero vector $\mathbf{0}$, or an infinite number of solutions.

<u>Example 1.4.6</u>

Solve the matrix system $\mathbf{A} \cdot \mathbf{x} = \mathbf{0}$ given

$$\mathbf{A} = \begin{bmatrix} 1 & 2 & 3 \\ 4 & 4 & 2 \\ 2 & 1 & 2 \end{bmatrix}$$

From <u>Example 1.4.5.</u> we can continue from the last elementary row operation at

$$\begin{bmatrix} 1 & 2 & 3 \\ 0 & 1 & 5/2 \\ 0 & 0 & 1 \end{bmatrix} \begin{pmatrix} x_1 \\ x_2 \\ x_3 \end{pmatrix} = \begin{pmatrix} 0 \\ 0 \\ 0 \end{pmatrix}$$

as follows:

$$\begin{bmatrix} 1 & 2 & 0 \\ 0 & 1 & 0 \\ 0 & 0 & 1 \end{bmatrix} \begin{pmatrix} x_1 \\ x_2 \\ x_3 \end{pmatrix} = \begin{pmatrix} 0 \\ 0 \\ 0 \end{pmatrix} \quad \begin{array}{l} \text{add } (-5/2) \text{ times} \\ \text{row 3 to row 2;} \\ \text{add } (-3) \text{ times row 3} \\ \text{to row 1} \end{array}$$

$$\begin{bmatrix} 1 & 0 & 0 \\ 0 & 1 & 0 \\ 0 & 0 & 1 \end{bmatrix} \begin{pmatrix} x_1 \\ x_2 \\ x_3 \end{pmatrix} = \begin{pmatrix} 0 \\ 0 \\ 0 \end{pmatrix} \quad \begin{array}{l} \text{add } (-2) \text{ times} \\ \text{row 2 to row 1;} \end{array}$$

The last two elementary row operations has the $\mathbf{A}$ matrix in **reduced row echelon form,** and the solution vector can be readily seen to be $(x_1, x_2, x_3) = (0, 0, 0)$. The technique of using elementary row operations on $\mathbf{A}$ (and $\mathbf{A}^*$) until an identity matrix is evolved (as when there is a unique solution vector) is called the **Gauss-Jordan** Elimination method.

Continuing with <u>Example 1.4.5,</u>

$$\begin{bmatrix} 1 & 2 & 3 \\ 0 & 1 & 5/2 \\ 0 & 0 & 1 \end{bmatrix} \begin{pmatrix} x_1 \\ x_2 \\ x_3 \end{pmatrix} = \begin{pmatrix} 1 \\ 1 \\ 0 \end{pmatrix}$$

as follows:

$$\begin{bmatrix} 1 & 2 & 0 \\ 0 & 1 & 0 \\ 0 & 0 & 1 \end{bmatrix} \begin{pmatrix} x_1 \\ x_2 \\ x_3 \end{pmatrix} = \begin{pmatrix} 1 \\ 1 \\ 0 \end{pmatrix}$$

add (-5/2) times row 3 to row 2; add (-3) times row 3 to row 1

$$\begin{bmatrix} 1 & 0 & 0 \\ 0 & 1 & 0 \\ 0 & 0 & 1 \end{bmatrix} \begin{pmatrix} x_1 \\ x_2 \\ x_3 \end{pmatrix} = \begin{pmatrix} -1 \\ 1 \\ 0 \end{pmatrix}$$

add (-2) times row 2 to row 1;

And $(x_1, x_2, x_3) = (-1, 1, 0)$.

Example 1.4.7

The Gauss-Jordan Elimination method can be extended to determine the inverse of nonsingular matrices. For an mxm matrix,

$$A_{mxm} \, A^{-1}_{mxm} = I_{mxm}$$

where I_{mxm} is the mxm identity matrix. In terms of column vectors,

$$A^{-1}_{mxm} = \begin{bmatrix} v_1, & v_2, \cdots v_m \end{bmatrix}$$

where each v_i is an mx1 column vector. Similarly,

$$I_{mxm} = \begin{bmatrix} W_1, & W_2, \cdots W_m \end{bmatrix}$$

where W_k is a mx1 zero column vector with a 1 in the k row. Then there are m matrix systems to solve of the form

$$A_{mxm} \, v_{i_{mx1}} = W_{i_{mx1}} \; ; \; i=1,2,\cdots,m$$

But each solution effort involves the same sequence of elementary row operations in transforming A_{mxm} into I_{mxm} by the Gauss-Jordan technique. Consequently, all m problems can be solved in tandem by using matrix A augmented with the m column vectors, W_i.

For A from Example 1.4.6, the augmented matrix system to find A^{-1} is written as

$$\begin{bmatrix} 1 & 2 & 3 & 1 & 0 & 0 \\ 4 & 4 & 2 & 0 & 1 & 0 \\ 2 & 1 & 2 & 0 & 0 & 1 \end{bmatrix}$$

Using the same sequence of elementary row operations as Example 1.4.6, the augmented matrix becomes

$$\begin{bmatrix} 1 & 0 & 0 & -6/14 & 1/14 & 8/14 \\ 0 & 1 & 0 & 4/14 & 4/14 & -10/14 \\ 0 & 0 & 1 & 4/14 & -3/14 & 4/14 \end{bmatrix}$$

and

$$A^{-1} = \frac{1}{14} \begin{bmatrix} -6 & 1 & 8 \\ 4 & 4 & -10 \\ 4 & -3 & 4 \end{bmatrix}$$

Note that the unique solution vector x to Example 1.4.5, is

$$x = A^{-1}b = \frac{1}{14} \begin{bmatrix} -6 & 1 & 8 \\ 4 & 4 & -10 \\ 4 & -3 & 4 \end{bmatrix} \begin{pmatrix} 1 \\ 0 \\ -1 \end{pmatrix} = \begin{pmatrix} -1 \\ 1 \\ 0 \end{pmatrix}$$

Example 1.4.8

Suppose one wishes to approximate the function f on the interval [a,b] using an interpolating polynomial of order 4, $p_4(x) = a_0 + a_1 x + a_2 x^2 + a_3 x^3 + a_4 x^4$. The approximation error, E, to be minimized on [a,b] is simply

$$E = \sum_{i=1}^{5} |p_4(x_i) - f(x_i)|$$

where the five x_i coordinates are interpolation points in [a,b], perhaps evenly spaced and including $x_0 = a$, $x_1 = a_1, \cdots$, $x_5 = b$.

Then the interpolation polynomial $p_4(x)$ is determined by solving the matrix system $Ax = b$ where

$$A = \begin{bmatrix} 1 & a & a^2 & a^3 & a^4 \\ 1 & x_2 & x_2^2 & x_2^3 & x_2^4 \\ 1 & x_3 & x_3^2 & x_3^3 & x_3^4 \\ 1 & x_4 & x_4^2 & x_4^3 & x_4^4 \\ 1 & b & b^2 & b^3 & b^4 \end{bmatrix} ; \quad x = \begin{pmatrix} a_0 \\ a_1 \\ a_2 \\ a_3 \\ a_4 \end{pmatrix} ; b = \begin{pmatrix} f(a) \\ f(x_2) \\ f(x_3) \\ f(x_4) \\ f(b) \end{pmatrix}$$

1.5. Metric Spaces

Let S be a set of objects (e.g., points x_j in the real number field **R**; functions x_j contained in the set of all continuous functions defined on the closed interval $[-\pi, \pi]$; integers x_j which are elements of the set of prime numbers; functions x_j which satisfy $\nabla^2 x_j = 0$ on the closed circle). Then S is said to be a **metric space** if a suitable metric or definition of the distance between each two elements of S is established. A **distance** function $d(x_i, x_j)$ qualifies as a **metric** on S if for x_i, x_j, x_k in S the following axioms are satisfied:

(i) $d(x_i, x_j)$ is a real number.

(ii) $d(x_i, x_j) \geq 0$.

(iii) $d(x_i, x_j) = 0$ only when $x_i = x_j$.

(iv) $d(x_i, x_j) = d(x_j, x_i)$.

(v) $d(x_i, x_j) \leq d(x_i, x_k) + d(x_k, x_j)$.

Typically, there are several candidates for a metric to use with a set of objects.

Example 1.5.1

Let $\hat{f}$ be an approximation of the function f over a domain $\mathcal{D}$ and suppose we are especially concerned in how accurately $\hat{f}$ approximates f at the two points x_1 and x_2 in $\mathcal{D}$. Let $e_1 = f(x_1) - \hat{f}(x_1)$ and $e_2 = f(x_2) - \hat{f}(x_2)$. Because we want $e(x) = f(x) - \hat{f}(x)$ to be as small as possible at the points x_1 and x_2, the measure of the approximation error is indicated by a value of distance between the ordered pairs (e_1, e_2) and $(0,0)$. A suitable metric is the Euclidean distance of $d_2((e_1, e_2), (0,0)) = (e_1^2 + e_2^2)^{1/2}$. Other choices for a metric include $d_\infty((e_1, e_2), (0,0)) = \max\{|e_1|, |e_2|\}$ or $d_1((e_1, e_2), (0,0)) = |e_1| + |e_2|$.

Notice that in this example, the set S is composed of two-dimensional vectors described by the ordered pair notation (x_1, x_2). Other choices for the approximation function $\hat{f}$ would typically result in different approximation errors (e_1, e_2) and the metric describes how well one approximation function succeeds in minimizing the approximation error at points x_1 and x_2 in $\mathcal{D}$.

Generally, the approximation error is evaluated at several points x_1, x_2,$\cdots$,x_n in $\mathcal{D}$ resulting in n–dimensional vectors (e_1, e_2,$\cdots$,e_n). The metric definitions considered in the above can be extended by $d_2(\underset{\sim}{e},\underset{\sim}{0}) = (e_1{}^2 + e_2{}^2 + \cdots + e_n{}^2)^{1/2}$; $d_\infty(\underset{\sim}{e},\underset{\sim}{0}) = \max\{|e_1|, |e_2|,\cdots, |e_n|\}$; $d_1(\underset{\sim}{e},\underset{\sim}{0}) = |e_1| + |e_2| + \cdots + |e_n|$ where vector notation is used to describe $\underset{\sim}{e} = (e_1, e_2\cdots,e_n)$ and accordingly $\underset{\sim}{0} = (0,0,\cdots,0)$. The subscript notation used in d_1, d_2 and d_∞ will be further discussed when ℓ_p spaces are examined; however, it can be noted that above expressions for d_1, d_2 and d_∞ are the ℓ_1, ℓ_2 and ℓ_∞ definitions for distance, respectively.

Once a metric has been established over the set S, several concepts regarding open and closed neighborhoods of a point $x \in S$ follow analogous to the familiar concepts in Euclidean geometry. The following definitions are important in the development of the necessary approximation theory.

DEFINITION 1.5.1 (Neighborhoods)

Let $\delta > 0$ and S be a metric space. Then a δ-neighborhood of the point $x \in S$ is the set of all points $y \in S$ such that $d(x,y) < \delta$. That is, $N_\delta(x) = \{y \in S: d(x,y) < \delta\}$. A closed δ-neighborhood of $x \in S$ is defined as $\bar{N}_\delta(x) = \{y \in S: d(x,y) \leq \delta\}$. A deleted neighborhood of $x \in S$ is defined as the set of all points $y \in S$ such that $0 < d(x,y) < \delta$.

What a neighborhood looks like depends on the metric used to describe the distance between points. For example, in the usual two-dimensional Euclidean space with vectors described by the Cartesian plane coordinates (x,y), closed δ-neighborhoods of the origin having coordinates (0,0) is shown in Figure 1.1 for the ℓ_1, ℓ_2, and ℓ_∞ definitions of distance, respectively.

DEFINITION 1.5.2 (Converging Sequence)

A sequence of vectors x_1, x_2,$\cdots$ in a metric space S is said to **converge** to a point x^* if the distance between x_n and x^* approaches zero as $n \to \infty$. That is, $x_n \to x^*$ if as $n \to \infty$, $d(x_n,x^*) \to 0$.

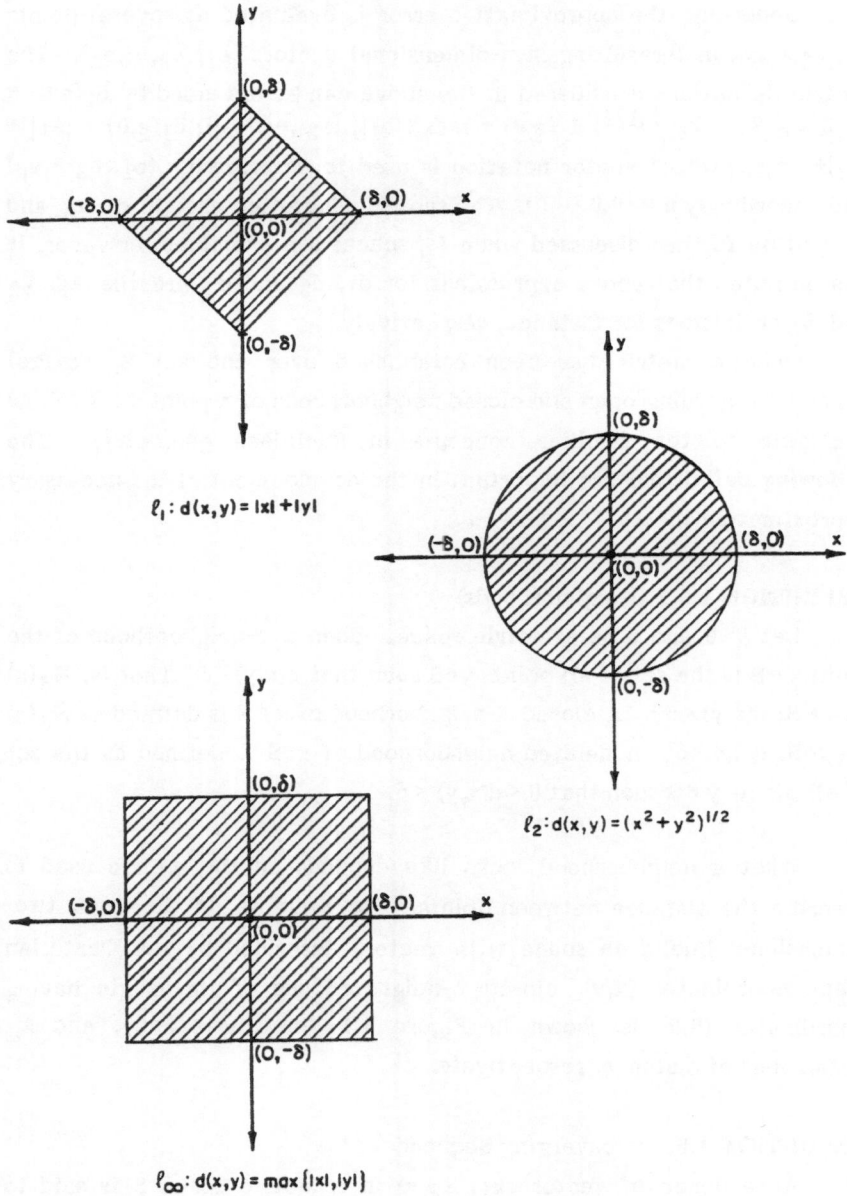

$\ell_1 : d(x,y) = |x| + |y|$

$\ell_2 : d(x,y) = (x^2 + y^2)^{1/2}$

$\ell_\infty : d(x,y) = \max\{|x|,|y|\}$

Fig. 1.1. Closed δ-Neighborhoods of Point $(0,0)$ for ℓ_1, ℓ_2 and ℓ_∞ Definitions of Distance.

Example 1.5.2

Let $x_n = n^2 e^{-n}$ where n is a positive integer. Then the sequence $\{x_n\}$ tends to the limiting value of $x^* = 0$ as $n \to \infty$; that is $\lim\limits_{n \to \infty} x_n = 0$. One method of proving that the above limit statement is true is to use an epsilon (ε) proof where it is shown that for any $\varepsilon > 0$ there exists a positive integer N (depending on ε) such that $d(x_n - x^*) < \varepsilon$ when $n \geq N$. For the subject sequence, let $d(x_n - x^*) = |x_n - x^*| = |n^2 e^{-n}| = n^2 e^{-n}$ for $n \geq 1$. If $\varepsilon = 0.001$, then a suitable N is $N = 12$. If various values of $n \leq 12$ are substituted into the formula for x_n it is seen that indeed $|x_n - x^*| < 0.001$. Should $\varepsilon = 10^{-10}$, a suitable N is $N = 30$. Because there exists an N for each ε, it is shown that $x_n \to x^*$ as $n \to \infty$. A more rigorous proof is to note that the Taylor series of $e^n = 1 + n + n^2/2! + n^3/3! + \cdots$ and therefore $n^2 e^{-n} < n^2/(1 + n + n^2/2 + n^3/6) < 6/n$. Then for any $\varepsilon > 0$, setting $N > 6/\varepsilon$ guarantees that $n^2 e^{-n} < \varepsilon$. Thus, for every $\varepsilon > 0$ there exists an $N = N(\varepsilon)$, proving that $n^2 e^{-n} \to 0$ as $n \to \infty$.

Example 1.5.3

Suppose now that the sequence being studied $\{x_n\}$ is composed of two-dimensional vectors defined by $x_n = (ne^{-n}, n^2 e^{-n})$. It is seen that $x_n \to x^* = (0,0)$ as $n \to \infty$. To prove this statement, it is noted that $ne^{-n} < 2/n$ and $n^2 e^{-n} < 6/n$. Let the metric be the ℓ_∞ distance $d_\infty(x_n, x_m) = \max\{|ne^{-n} - me^{-m}|, |n^2 e^{-n} - m^2 e^{-m}|\}$. Then $d_\infty(x_n, x^*) = \max\{ne^{-n}, n^2 e^{-n}\} = n^2 e^{-n}$. But from the previous one-dimensional example, $n^2 e^{-n} < 6/\varepsilon$ implies $N(\varepsilon) \geq 6/\varepsilon$ for all $\varepsilon > 0$. Thus for every $\varepsilon > 0$ there exists an N such that $n \geq N(\varepsilon) \to d_\infty(x_n, x^*) < \varepsilon$. Thus, $(ne^{-n}, n^2 e^{-n}) \to (0,0)$ as $n \to \infty$.

Suppose that the ℓ_1 distance of $d_1(x_n, x_m) = |ne^{-n} - me^{-m}| + |n^2 e^{-n} - m^2 e^{-m}|$ is used as the metric. Then to prove the limit statement we must show that as $n \to \infty$, $d_1(x_n, x^*) \to 0$. But $d_1(x_n, x^*) = ne^{-n} + n^2 e^{-n} \leq 2n^2 e^{-n} < 12/n$. Hence, for every $\varepsilon > 0$ there exists an $N(\varepsilon) = 12/\varepsilon$ such that $n \geq N(\varepsilon) \to d_1(x_n, x^*) < \varepsilon$. Again, $(ne^{-n}, n^2 e^{-n}) \to (0,0)$ as $n \to \infty$.

Using the ℓ_2 distance, and $x_m \to 0$, $d_2(x_n, x_m) = ((ne^{-n} - me^{-m})^2 + (n^2 e^{-n} - m^2 e^{-m})^2)^{1/2} < (2(n^2 e^{-n})^2)^{1/2} < \sqrt{2}\, n^2 e^{-n} < 6\sqrt{2}/n$. Then for every $\varepsilon > 0$ there exists an $N(\varepsilon) \geq 6\sqrt{2}/\varepsilon$ such that $n \geq N(\varepsilon) \to d_2(x_n, x^*) < \varepsilon$.

Example 1.5.4

As a final illustration, consider a sequence of dimension 100 vectors defined by $x_n = (ne^{-n}, n^2e^{-n}, \cdots, n^{100}e^{-n})$. Then it can be shown that $x_n \to x^* = (0, 0, \cdots, 0)$ as $n \to \infty$ by noting that $n^{100}e^{-n} < 101!/n$. Thus for the ℓ_1 distance as a metric, $d_1(x_n, x^*) = \sum_{j=1}^{100} n^j e^{-n} < 100n^{100}e^{-n} < (100)(101!/n)$.

Letting $N(\varepsilon) \geq (100)(101!)/\varepsilon$ shows the existence of $N(\varepsilon)$ for every $\varepsilon > 0$ in the limit criteria. For the ℓ_∞ metric, $N(\varepsilon) > 101!/\varepsilon$. For the ℓ_2 metric, $N(\varepsilon) = (10)(101!)/\varepsilon$.

The above examples suggest that the limit of a sequence is independent of the choice of the metric, and that multidimensional vectors follow the same limit criteria that is understood for the basic one-dimensional case.

We are interested in the above two concepts because the Best Approximation Method will be using multidimensional vectors to represent the approximation function components, and the best approximation will be developed using the ℓ_2 definition for distance as the metric. That is, the Best Approximation Method will determine the best approximation function (from a set of basis functions) in an ℓ_2 norm in that the ℓ_2 error in satisfying the given linear operator relationship and the boundary and initial conditions will be a minimum.

1.6. Linear Spaces

The subject of linear spaces and linear operators is fundamental to the development of the theory utilized in methods of numerical approximation.

DEFINITION 1.6.1 (Linear Space or Vector Space)

Let $S = \{x, y, z, \cdots\}$ be a nonempty set of elements or vectors and $F = \{\alpha, \beta, \gamma, \cdots\}$ be a scalar field (e.g., the real number field, $\mathbb{R}$). Also let there be defined the operations of addition between every two elements of S and scalar multiplication between every element of S and every element of F such that for all $x, y, z \in S$ and $\alpha, \beta \in F$:

(i) $x + y \in S$.

(ii) $\alpha x \in S$.

(iii) $x + y = y + x$.

(iv) $(x + y) + z = x + (y + z)$.

(v)   There exists a zero element in S, designated by θ, such that for $x \in S$, $x + \theta = x$.

(vi)   For each $x \in S$, there also exists in S the negative of x (noted as $-x$) such that $x + (-x) = \theta$.

(vii) $\alpha(\beta x) = (\alpha \beta) x$

(viii) $\alpha(x + y) = \alpha x + \alpha y$

(ix) $(\alpha + \beta) x = \alpha x + \beta x$

(x) $1x = x$

Then S is said to be a **linear space** over the field F. If F is the real number field, **R**, then S is a **real linear space**. The real linear space, S, is the focus for the development of the Best Approximation Method. If F is the field of complex numbers, then S is a **complex linear space**.

Note that the operations of element addition and scalar multiplication are similar to computer programs in that $f + g = h$ implies two inputs of vectors with an output of a vector which is also in S, and $\alpha f = g$ implies a scalar input and a vector input with a vector output. It is oftentimes useful to designate element addition in S by $[+,S]$, and scalar multiplication by $[\cdot,S,F]$. Thus element addition would be noted as $f[+,S]g$, and scalar multiplication by $\lambda[\cdot,S,F]f$ for $f, g \in S$, $\lambda \in F$.

DEFINITION 1.6.2 (Linear Combination)

Let $x_1, x_2, \cdots, x_n$ be elements of S and $\alpha_1, \alpha_2, \cdots, \alpha_n$ be elements of **R**. The sum $\alpha_1 x_1 + \alpha_2 x_2 + \cdots + \alpha_n x_n$ is called a **linear combination** of the x_j.

DEFINITION 1.6.3 (Linear Independence)

Let $x_1, x_2, \cdots, x_n$ and $\alpha_1, \alpha_2, \cdots, \alpha_n$ be elements of S and **R**, respectively. Then the x_j are called **linearly independent** if and only if the sum $\alpha_1 x_1 + \alpha_2 x_2 + \cdots + \alpha_n x_n = \theta$ implies each scalar α_j is zero. If there exists a set $\alpha_1, \alpha_2, \cdots, \alpha_n \in \mathbf{R}$ such that not all of these elements are zero and yet the subject sum is θ, then the set of x_j are said to be **linear dependent**.

DEFINITION 1.6.4 (Dimension of the Linear Space, S)

Let S be a linear space. Suppose that there exist n elements x_1, x_2,$\cdots$,x_n in S which are linearly independent, and every set of $n + 1$ elements in S is linearly dependent. Then the **dimension** of S is said to be n, noted by Dim (S) = n. If there exist m linearly independent elements in S for every integer $m > 0$, then Dim (S) = ∞.

DEFINITION 1.6.5 (Basis for a Linear Space, S)

Let x_1, x_2,$\cdots$,x_n be a linearly independent set of elements in S such that each x in S can be written as a linear combination of the x_j, $j = 1,2,\cdots,n$. Then the x_j are said to form a **basis** of S. The set $\{x_j, j = 1,2,\cdots n\}$ is said to **span** the linear space, S. (Hereafter, it is understood that the given index follows the positive integers.)

Example 1.6.1

An example of a linear space is the set $\mathbf{R}^m$ of all m-dimensional vectors $(x_1, x_2,\cdots,x_m)$ where each $x_j \in \mathbf{R}$. Addition, subtraction, and scalar multiplication are all defined in the usual way. The zero vector is defined by $\theta = (0,0,\cdots,0)$. It follows immediately that Dim $(\mathbf{R}^m) = m$ and a basis for $\mathbf{R}^m$ is the m vectors $\{x_k\}$ where x_k is the vector with 1 as its kth component and 0 for its remaining components.

Example 1.6.2

An important example of a linear space is the set of all functions which are continuous over the closed interval $[a,b]$. This space is noted by $C[a,b]$ and satisfies all the properties of Definition 1.6.1. For two vectors f_1 and f_2 in $C[a,b]$, addition is defined by $(f_1 + f_2)(x) = f_1(x) + f_2(x)$ for each point x in $[a,b]$, and scalar multiplication is defined for $\lambda \in \mathbf{R}$ by $\lambda f_1 = \lambda f_1(x)$. The zero vector is defined by the continuous function $f_1(x) = 0$, for x in $[a,b]$. Subtraction is defined by $(f_1-f_2)(x) = f_1(x) - f_2(x)$. It is noted that the addition and subtraction of f_1 and f_2 is another continuous function, and the scalar multiplication of f_1 is a continuous function; that is, $(f_1 + f_2) \in C[a,b]$, $(f_1 - f_2) \in C[a,b]$, and $\lambda f_1 \in C[a,b]$.

A linear combination of vectors $f_j \in C[a,b]$ is a continuous function $f = \sum\limits_{j=1}^{m} \lambda_j f_j \in C[a,b]$.

1.7. Normed Linear Spaces

A generalization of the concept of distance between two vectors of a linear space is the norm.

Definition 1.7.1 (Norm)

A **norm** is a real valued function $||x||$ defined for each $x \in S$ with the following properties:

(i) $||x|| \geq 0$

(ii) $||\lambda x|| = |\lambda| \, ||x||$ for $\lambda \in \mathbf{R}$

(iii) $||x + y|| \leq ||x|| + ||y||$ for all $x, y \in S$ (triangle inequality)

(iv) $||x|| = 0$ implies $x = \theta$

A linear space with an associated norm is called a **normed linear space**. A normed linear space is also a metric space and, therefore, the properties and concepts embodied in metric spaces (e.g., neighborhoods) also apply in a normed linear space.

Example 1.7.1

In the vector space $\mathbf{R}^m$, a well used norm is the ℓ_2 norm where for $x, y, \in \mathbf{R}^m$, $||x - y|| = ((x_1 - y_1)^2 + (x_2 - y_2)^2 + \cdots + (x_m - y_m)^2)^{1/2}$. It is seen that the properties of Definition 1.7.1 are satisfied:

(i) $||x|| \geq 0$

(ii) $||\lambda x|| = ((\lambda x_1)^2 + \cdots + (\lambda x_m)^2)^{1/2}$
$$= (\lambda^2(x_1^2 + x_2^2 + \cdots + x_m^2))^{1/2}$$
$$= |\lambda| \, (x_1^2 + x_2^2 + \cdots + x_m^2)^{1/2} = |\lambda| \, ||x||$$

(iii) $||x + y|| \leq ||x|| + ||y||$. In Euclidian one, two, and three-dimensional space where the norm is defined to be the usual Euclidian distance, the triangle inequality follows. For higher dimensions, or for other norms, Minkowski's inequality must be used as a proof.

(iv) $||x|| = 0$ implies $x_1^2 = x_2^2 = \cdots x_m^2 = 0$, and $x = \theta$.

Example 1.7.2.

In $C[a,b]$, a norm $||f||$ where $f \in C[a,b]$ can be defined as $||f|| = \max \{|f(x)| : x \in [a,b]\}$. Consequently for f_1 and f_2 in $C[a,b]$, $||f_1 - f_2|| = \max \{|f_1(x) - f_2(x)| : x \in [a,b]\}$. Figure 1.2 shows the definition of $||f_1 - f_2||$. The norm here is called the maximum norm.

Let S be a normed linear space and let $\{x_n\}$ be a sequence of vectors such that $x_n \in S$ for each n. Suppose that $x_n \to x^*$ as $n \to \infty$. Then if $x^* \in S$, the sequence $\{x_n\}$ converges to a vector x^* in S. However, oftentimes a sequence can converge to a vector x^* but x^* is not an element of S. This can be illustrated by the following examples:

Example 1.7.3

Let S be the open interval (0,1) with the norm $||x|| = |x|$ for $x \in S$. Let $\{x_n\}$ be the sequence of vectors (points) defined by $x_n = \frac{1}{n}$ for each $n \geq 1$. Then $x_n \to x^* = 0$ as $n \to \infty$. But $0 \notin S$, hence $x_n \to x^* \notin S$ as $n \to \infty$.

Example 1.7.4

Let S be the set of all rational numbers, Q (i.e., the set of all real numbers q such that q = m/n where m and n are integers). Let T be the set of all irrational numbers (i.e., $T = R-Q$). Let t be an element of T and let x_n be the first n digits of the decimal expansion of t. Then each $x_n \in S$, and as $n \to \infty$ $x_n \to t \notin S$. Now consider $\{x_n\} = \{1, 1+1, 1+1+ \frac{1}{2!} \cdots\}$ where $x_n = \sum_{j=0}^{n} 1/j!$. Then $x_n \to x^* = e$ as $n \to \infty$ (the term x_n is the first n terms of the Taylor series for e^x at x = 1). Each term x_n is in Q(i.e., S) and therefore $\{x_n\}$ is in Q. But $x_n \to x^* = e$ as $n \to \infty$ where $e \in T$ and $e \notin Q$. Thus, $x_n \to x^* \notin S$ as $n \to \infty$.

Example 1.7.5

As in Example 1.7.4 let S = Q and T = R - Q. Then convergent sequences $\{x_n\}$ in S can be constructed which converge to vectors $x^* \in T$ for each $t \in T$; that is, the space Q is **incomplete**. In contrast, had S = R, then each convergent sequence $\{x_n\}$ in R converges to a point $x^* \in R$. The space R is said to be **complete**.

The definition of a convergent sequence given in Definition 1.5.2 is too general for practical purposes: the limit of the sequence, x^*, must be known to prove $\{x_n\} \to x^*$ as $n \to \infty$.

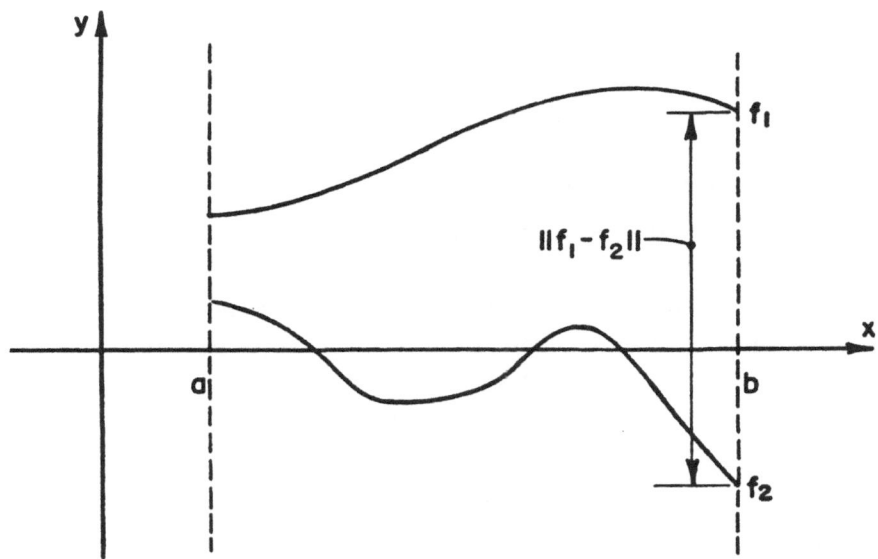

Fig. 1.2.   Definition of $||f_1 - f_2||$ for f_1, $f_2 \in C\ [a,b]$

Another approach to evaluating whether a sequence converges is to show that for every $\varepsilon > 0$ there exists a positive integer $N(\varepsilon)$ such that $||x_n - x_{n+p}|| < \varepsilon$ for $n \geq N(\varepsilon)$ and $p = 1,2,\cdots$. That is, for any $n \geq N(\varepsilon)$, all of the remaining terms x_j of the sequence ($j>n$) lie in the closed ε-neighborhood $\bar{N}_\varepsilon(x_n)$. As $\varepsilon \to 0$ the associated $N(\varepsilon)$ get larger, resulting in a parallel sequence of closed ε-neighborhoods where each successive $\bar{N}_{\varepsilon k+1}(x_n)$ is contained in the previous $\bar{N}_{\varepsilon k}(x_n)$ for $\varepsilon_k > \varepsilon_{k+1}$. This sequence of nested closed neighborhoods focuses about the limit vector x^* as $N(\varepsilon) \to \infty$.

A sequence $\{x_n\}$ which satisfies the above criteria for convergence is called a **Cauchy sequence.**

It is noted that whether the limit point x^* is an element of S is not established by simply showing the sequence $\{x_n\}$ is Cauchy.

A normed linear space S where any Cauchy sequence $\{x_n\}$ has its limit x^* contained in S is said to be **complete**. A complete normed linear space is called a **Banach space.**

Example 1.7.6

The normed linear space R^m is a Banach space. This can be shown by noting that a sequence $\{x_n\}$ in R^m is composed of vectors of the form $(x_1, x_2, \cdots, x_n)$ where each coordinate x_j of the vectors is also a Cauchy sequence which converges to its own limit point, x_j^* (the space R is complete). Thus each $\{x_j\} \to x_j^*$ as $n \to \infty$. Thus the vectors $\{x_n\} \to (x_1^*, x_2^*, \cdots, x_m^*) \in R^m$ as $n \to \infty$. Hence, each Cauchy sequence in R^m converges to a limit which is an element of R^m. Thus R^m is a Banach space.

Example 1.7.7

$C[a,b]$ with the maximum norm is a Banach space. Let $\{f_n\}$ be a Cauchy sequence in $C[a,b]$. Using the norm defined in Example 1.7.4 it is seen that for $\varepsilon > 0$ there exists an $N(\varepsilon)$ such that $||f_n - f_{n+p}|| < \varepsilon$ for $n \geq N(\varepsilon)$ and $p = 1,2,\cdots$. Thus, $|f_n(x) - f_{n+p}(x)| < \varepsilon$ for $x \in [a,b]$. Let $x_0 \in [a,b]$. Then $|f_n(x_0) - f_{n+p}(x_0)| < \varepsilon$ for $n \geq N(\varepsilon)$ and $p = 1,2,\cdots$. That is $\{f_n(x_0)\}$ is a Cauchy sequence in R and $f_n(x_0) \to f_n^*(x_0) \in R$ as $n \to \infty$. Thus a limit $f_n^*(x)$ exists for $\{f_n(x)\}$ for all $x \in [a,b]$. Now we must see if $f_n^*(x)$ is continuous (i.e., show $f_n^*(x) \in C[a,b]$). To show $f_n^*(x)$ is continuous, we must show that a small δ-change in x (in [a,b]) results in a small change in $f_n^*(x)$. But $|f_n^*(x+\delta) - f_n^*(x)| = |f_n^*(x+\delta) - f_n(x+\delta) + f_n(x+\delta) - f_n(x) + f_n(x) - f_n^*(x)| \leq |f_n^*(x+\delta) - f_n(x+\delta)| + |f_n(x+\delta) - f_n(x)| + |f_n(x) - f_n^*(x)|$. The middle term $|f_n(x+\delta) - f_n(x)|$ reflects the continuity of the terms f_n in the sequence and can be made arbitrarily small (say less than $\varepsilon/3$) by choosing δ small enough. The other two terms reflect the convergence of the two Cauchy sequences $\{f_n^*(x+\delta)\}$ and $\{f_n^*(x)\}$. Choosing $\varepsilon = \varepsilon^1/3$ determines $N(\varepsilon^1/3)$ such that $|f_n^*(x + \delta) - f_n(x + \delta)| + |f_n(x) - f_n^*(x)| < \varepsilon^1/3 + \varepsilon^1/3 = 2\varepsilon^1/3$. Thus, for any $\varepsilon^1 > 0$ there exists a $\delta > 0$ (and an $N(\varepsilon^1)$) such that $|f_n^*(x+\delta) - f_n^*(x)| < 3\varepsilon^1 = \varepsilon$. Since f_n^* is continuous, $f_n^* \in C[a,b]$. Thus a Cauchy sequence in $C[a,b]$ contains a limit in $C[a,b]$. Therefore $C[a,b]$ is a Banach space.

Example 1.7.8

The space $C[a,b]$ with the norm $||f|| = (\int_a^b f^2(x)dx)^{1/2}$ is not complete. This can be shown by example. Let [a,b] = [-1,1] and the sequence $\{f_n\}$ be in $C[-1,1]$ where

$$f_n(x) = \begin{cases} -1, & -1 \leq x \leq -1/n \\ nx, & -1/n \leq x \leq 1/n \\ 1, & \frac{1}{n} \leq x \leq 1 \end{cases}$$

Using the defined norm $||f||$, $||f_n - f_{n+p}|| <$

$$(2 \int_0^{1/n} (1 - nx)^2 dx)^{1/2} = (2 \int_0^{1/n} (1 - 2nx + n^2x^2)dx)^{1/2}$$

$$= (1/6n)^{1/2}$$

Thus letting $N(\varepsilon) \geq 1/6\varepsilon^2$ we see gives that for every $\varepsilon > 0$ there exists an $N(\varepsilon)$ such that $||f_n - f_{n+p}|| < \varepsilon$ for $n \geq N(\varepsilon)$ and $p = 1, 2, \cdots$. Thus, $\{f_n\}$ is a Cauchy sequence using the defined norm. However, as $n \to \infty$, $f_n \to f^*$ where

$$f^*(x) = \begin{cases} -1, & -1 \leq x \leq 0 \\ 0, & x = 0 \\ 1, & 0 \leq x \leq 1 \end{cases}$$

Thus $f_n \to f^*$ as $n \to \infty$, but f^* is not continuous over [-1,1].

DEFINITION 1.7.2 (inner product)

Let S be a real linear space. The real function (,) is called an **inner product** in S if for any elements x_1, x_2, x_3 in S, (x_1, x_2) satisfies

(i) $(x_1,x_1) \geq 0$

(ii) $(x_1,x_1) = 0$ implies $x_1 = \theta \in S$

(iii) $(x_1,x_2) = (x_2,x_1)$

(iv) for $\lambda \in \mathbf{R}$, $(\lambda x_1,x_2) = \lambda(x_1,x_2)$

(v) $(x_1 + x_2, x_3) = (x_1,x_3) + (x_2,x_3)$

Given an inner product (,) on the real linear space S, a norm of $x_1 \in S$ is

$$||x_1|| = (x_1,x_1)^{1/2}$$

The real linear space S together with (,) is called a **real inner product space**.

Example 1.7.9

An important inner product on $S = C[a,b]$ for f and g in S is

$$(f,g) = \int_a^b f(x)g(x) \, dx$$

where

$$(f,f) = \int_a^b f^2(x) \, dx$$

and

$$|| f || = (\int_a^b f^2(x) \, dx)^{1/2}$$

Definition 1.7.3 (Hilbert space)

A complete inner product space is called a **Hilbert** space.

Theorem 1.7.1 (Schwarz Inequality)

Let S be a Hilbert space, with f and g in S. Then

$$|(f,g)| \leq ||f|| \, ||g||$$

PROOF: Let f and g be elements of S where neither f nor g are the zero element $\theta \in S$ (otherwise, the proof is trivial). Let $\lambda \in \mathbf{R}$.

CLAIM	SUBSTANTIATION
(1.) $(f+\lambda g, f+\lambda g) \geq 0$	property (i) of Definition 1.7.2
(2.) $(f,f)+2\lambda(f,g)+\lambda^2(g,g) \geq 0$	properties (iii), (iv), (v)
(3.) $(g,g) \neq 0$	$g \neq \theta \in S$; property (ii)
(4.) define $\lambda = -(f,g)/(g,g)$	λ can be any real number
(5.) $(f,f)-(f,g)^2/(g,g) \geq 0$	Substitution of λ in (4) into (2)
(6.) $(f,f) \geq (f,g)^2/(g,g)$	addition property of inequality
(7.) $(f,g)^2 \leq (f,f)(g,g)$	$g \neq \theta$, thus $(g,g) > 0$ by properties (i), (ii)
(8.) $|(f,g)| \leq ||f|| \, ||g||$	$(f,f) = ||f||^2$; $||f|| > 0$.

Example 1.7.10

$S = \mathbf{R}^n$ is the linear space of column vectors (or row vectors) of dimension n with each component in $\mathbf{R}$. A suitable inner-product in S is (,) being the usual dot product. That is, for **u** and **v** in S, and u_i the i-row value in **u**,

$$(\mathbf{u}, \mathbf{v}) = \mathbf{u} \cdot \mathbf{v} = \sum_{i=1}^{n} u_i v_i$$

Because each Cauchy sequence in $\mathbf{R}^n$ has a limit point in $\mathbf{R}^n$, S is complete, and S is a Hilbert space.

1.8. Approximations

A great many problems in computational mechanics involve the development of an approximation of the form

$$\hat{\phi} = \sum_{i=1}^{n} \lambda_i f_i \tag{1.8.1}$$

where the f_i are functions, vectors, or elements chosen from modeling experience, and the λ_i are real constants to be computed. The problem setting is to have a relationship or operator O with auxilliary or boundary conditions such that, in general terms, a unique element f solves

$$Of = g \qquad (1.8.2)$$

where g is a known element. This **inverse problem** setting is a focus of computational mechanics, and includes classes of problems such as solving partial differential equations, integral equations, among others.

The particular strategy in choosing the element f_i of (1.8.1) depends upon the problem setting. Notwithstanding, the approximation $\hat{\phi}$ is used to approximate f by

$$O\hat{\phi} = g \qquad (1.8.3)$$

to directly or indirectly compute real constants, λ_i in (1.8.1). In order to obtain a well-posed problem in solving for λ_i values, the f_i elements are chosen to be linearly independent. Thus the n elements f_i form a basis for a linear space S with dim $S = n$. Unfortunately, the solution f of (1.8.2) is seldom in S and cannot be derived, and $\hat{\phi}$ must be chosen such as to minimize some definition of approximation error. The choice of formula used to estimate error, E, usually impacts the computed λ_i values, and hence the resulting approximation $\hat{\phi}$ is influenced by the modelers preference for what type of E is being minimized.

Sometimes, the operator relationship can be "better" modeled by use of a particular set of elements f_i in (1.8.1) (i.e., **basis** or **trial** elements, vectors, or functions). Additionally, the chosen E definition can be used in our effort to lower E-values for a particular set of basis elements. Thus, the optimal $\hat{\phi}$ approximation depends upon the supplied basis elements and the chosen definition of error.

Example 1.8.1

Consider the second order ordinary homogeneous differential equation y" + y = 0 on domain $\Omega = [0, \pi/2]$ with auxilliary conditions $y(0) = 0$, $y(\pi/2) = 1$. Let the approximation $\hat{\phi}$ be

$$\hat{\phi}(x) = \lambda_1 x + \lambda_2 \sin x + \lambda_3 \cos x$$

Then $O\hat{\phi}(x) = \hat{\phi}''(x) + \hat{\phi}(x) = \lambda_1 x$.

Let ϕ be the yet unknown solution to the operator problem and define

$$E = \int_\Omega (O\hat{\phi} - O\phi)^2 \, d\Omega + (\hat{\phi}(0) - \phi(0))^2 + (\hat{\phi}(\pi/2) - \phi(\pi/2))^2$$

Then

$$E = \int_0^{\pi/2} (\lambda_1 x - 0)^2 \, dx + (\lambda_3 - 0)^2 + (\lambda_1 \frac{\pi}{2} + \lambda_2 - 1)^2$$

$$= \lambda_1^2 \, \pi^3/24 + \lambda_3^2 + (\lambda_1 \pi/2 + \lambda_2 - 1)^2$$

Because $E = E\,(\lambda_1, \lambda_2, \lambda_3)$, E is minimized by setting

$$\frac{\partial E}{\partial \lambda_1} = \lambda_1\,(\pi^2/12 + \pi/2) + \lambda_2 - 1 = 0$$

$$\frac{\partial E}{\partial \lambda_2} = \lambda_1\,\pi/2 + \lambda_2 - 1 = 0$$

$$\frac{\partial E}{\partial \lambda_3} = 2\lambda_3 = 0$$

In matrix form, E is minimized when

$$\begin{bmatrix} (\pi^2/12 + \pi/2) & 1 & 0 \\ \pi/2 & 1 & 0 \\ 0 & 0 & 1 \end{bmatrix} \begin{Bmatrix} \lambda_1 \\ \lambda_2 \\ \lambda_3 \end{Bmatrix} = \begin{Bmatrix} 1 \\ 1 \\ 0 \end{Bmatrix}$$

which has the unique solution $(\lambda_1, \lambda_2, \lambda_3) = (0, 1, 0)$. Thus the best approximation is $\hat{\phi}(x) = \sin x$, which is the solution to the operator problem.

Note that the $\{f_i\} = \{1, \sin x, \cos x\}$ span the linear space S where each element $s \in S$ is of the form $s = \lambda_1 + \lambda_2 \sin x + \lambda_3 \cos x$ for each $\lambda_i \in \mathbf{R}$. The dim S = 3. The solution to the example problem, ϕ, happened to be an element of S, and the approximation error is E = 0. (The solution is an element of the 2-dimension subspace of S spanned by $\{\sin x, \cos x\}$.

Example 1.8.2

Consider the space $S = \mathbf{R}^3$ and the operator relationship $\mathbf{Ax} = \mathbf{b}$ as defined in Example 1.4.5. The approximation space, S, has the basis $\{\mathbf{v_1}, \mathbf{v_2}, \mathbf{v_3}\}$ (as an example) where

$$\mathbf{v_1} = \begin{bmatrix} 1 \\ 0 \\ 1 \end{bmatrix} \quad ; \quad \mathbf{v_2} = \begin{bmatrix} 0 \\ 1 \\ 1 \end{bmatrix} \quad ; \quad \mathbf{v_3} = \begin{bmatrix} 1 \\ 2 \\ 0 \end{bmatrix}$$

The approximation is $\hat{\phi} = \lambda_1 \mathbf{v_1} + \lambda_2 \mathbf{v_2} + \lambda_3 \mathbf{v_3}$, and ϕ is the element that solves $\mathbf{Ax} = \mathbf{b}$, where $\mathbf{A}$ and $\mathbf{b}$ are defined in <u>Example 1.4.5</u>.

It is observed that

$$\begin{aligned} \mathbf{A}\hat{\phi} &= \lambda_1 \mathbf{A} \mathbf{v_1} + \lambda_2 \mathbf{A} \mathbf{v_2} + \lambda_3 \mathbf{A} \mathbf{v_3} = \mathbf{b} \\ &= \lambda_1 \begin{bmatrix} 4 \\ 6 \\ 4 \end{bmatrix} + \lambda_2 \begin{bmatrix} 5 \\ 6 \\ 3 \end{bmatrix} + \lambda_3 \begin{bmatrix} 5 \\ 12 \\ 4 \end{bmatrix} \end{aligned}$$

The approximation error is chosen to be

$$E = \sum_{i=1}^{3} (\hat{b}_i - b_i)^2 = (4\lambda_1 + 5\lambda_2 + 5\lambda_3 - 1)^2 \\ + (6\lambda_1 + 6\lambda_2 + 12\lambda_3 - 0)^2 + (4\lambda_1 + 3\lambda_2 + 4\lambda_3 + 1)^2$$

where b_i is the ith component of $\mathbf{b}$.

Then $E = E(\lambda_1, \lambda_2 \, \lambda_3)$ and is minimized by setting

$$\frac{\partial E}{\partial \lambda_1} = 136\lambda_1 + 136\lambda_2 + 216\lambda_3 = 0$$

$$\frac{\partial E}{\partial \lambda_2} = 136\lambda_1 + 140\lambda_2 + 218\lambda_3 - 4 = 0$$

$$\frac{\partial E}{\partial \lambda_3} = 216\lambda_1 + 218\lambda_2 + 270\lambda_3 - 2 = 0$$

In matrix form, E is minimized when

$$\begin{bmatrix} 136 & 136 & 216 \\ 136 & 140 & 218 \\ 216 & 218 & 270 \end{bmatrix} \begin{bmatrix} \lambda_1 \\ \lambda_2 \\ \lambda_3 \end{bmatrix} = \begin{bmatrix} 0 \\ 4 \\ 2 \end{bmatrix}$$

which gives $(\lambda_1, \lambda_2, \lambda_3) = (-1, 1, 0)$.

And $\hat{\phi} = \lambda_1 \mathbf{v_1} + \lambda_2 \mathbf{v_2} + \lambda_3 \mathbf{v_3} = (-1, 1, 0)$ is also the solution to the $\mathbf{Ax} = \mathbf{b}$ problem.

Example 1.8.3

In this example, we want to find the "best" constant λ to use in approximating x^2 on $\Omega = [0,1]$. The first choice for the definition of error is E_1 where

$$E_1 = \int_0^1 |\lambda - x^2| \, dx = \int_0^{\sqrt{\lambda}} (\lambda - x^2) \, dx + \int_{\sqrt{\lambda}}^1 (x^2 - \lambda) \, dx$$

Integrating, $E_1 = E_1(\lambda) = \frac{4}{3}\lambda^{3/2} - \lambda + 1/3$
and E_1 is minimized when

$$\frac{\partial E_1}{\partial \lambda} = 2\lambda^{1/2} - 1 = 0, \text{ or } \lambda = \frac{1}{4}.$$

Note that $\frac{\partial^2 E_1}{\partial \lambda^2} = \lambda^{-1/2}$ which is positive at $\lambda = 1/4$, hence $\lambda = 1/4$ minimizes E_1.

Now for comparison purposes use E_2 where

$$E_2 = (\int_0^1 |\lambda - x^2|^2 \, dx)^{1/2}, \text{ or } E_2^2 = \int_0^1 (\lambda - x^2)^2 \, dx$$

Integrating, $E_2(\lambda) = \lambda^2 - \frac{2}{3}\lambda + \frac{1}{5}$

E_2 is minimum when

$$\frac{\partial E_2}{\partial \lambda} = 2\lambda - \frac{2}{3} = 0, \text{ or when } \lambda = \frac{1}{3}$$

Note, $\frac{\partial^2 E_2}{\partial \lambda^2} = 2$ is positive.

Thus the determining of the "best" approximation λ used to approximate x^2 on $[0,1]$ depends on the choice of error definition.

Example 1.8.4

Consider the real linear space S with basis $\{v_1, v_2\}$ where $v_1 = (1,2,3,4)^T$, $v_2 = (2,0,0,1)^T$, where S is a linear subspace of $\mathbf{R}^4$.
Let $g = (1,2,1,2)^T$. The problem to be solved is finding the best approximation $\hat{\phi}$ S of g. Then

$$\hat{\phi} = \lambda_1 v_1 + \lambda_2 v_2 = \hat{g}$$

The error definition used is

$$E = \sum_{i=1}^{4} (\hat{g}_i - g_i)^2$$

where g_i is the ith component of **g**. Then

$$E = E(\lambda_1, \lambda_2) = (\lambda_1 + 2\lambda_2 - 1)^2 + (2\lambda_1 - 2)^2$$
$$+ (3\lambda_1 - 1)^2 + (4\lambda_1 + \lambda_2 - 2)^2$$

and E is minimum when

$$\frac{\partial E}{\partial \lambda_1} = 60\lambda_1 + 12\lambda_2 - 32 = 0$$

$$\frac{\partial E}{\partial \lambda_2} = 12\lambda_1 + 10\lambda_2 - 8 = 0$$

or, in matrix form

$$\begin{bmatrix} 60 & 12 \\ 12 & 10 \end{bmatrix} \begin{pmatrix} \lambda_1 \\ \lambda_2 \end{pmatrix} = \begin{pmatrix} 32 \\ 8 \end{pmatrix}$$

and $(\lambda_1, \lambda_2) = (28/57, 4/19)$.

The "best" approximation of **g** is

$$\hat{\phi} = \lambda_1 \mathbf{v_1} + \lambda_2 \mathbf{v_2}$$

$$= \frac{28}{57} \begin{pmatrix} 1 \\ 2 \\ 3 \\ 4 \end{pmatrix} + \frac{12}{57} \begin{pmatrix} 2 \\ 0 \\ 0 \\ 1 \end{pmatrix} = \begin{pmatrix} 52/57 \\ 56/57 \\ 84/57 \\ 124/57 \end{pmatrix}$$

It is noted that since $\hat{\phi} \neq \mathbf{g}$, then $\mathbf{g} \notin S$. This example demonstrates approximating an element not in the linear space spanned by the basis elements. Should two more linearly independent elements $\mathbf{v_3}$ and $\mathbf{v_4}$ be added to the original basis, then $S = \mathbb{R}^4$ and $\mathbf{g} \in S$.

CHAPTER 2
INTEGRATION THEORY

2.0. Introduction

Before developing further the concepts of generalized Fourier series, Hilbert spaces, and error analysis of computational methods, a brief presentation of Lebesgue integration theory is needed for completeness. These integration concepts find use in development of error bounds, and also are necessary in providing limits of sequences of approximations.   In this Chapter, only a brief survey of the more important concepts of Lebesgue integration theory is reviewed, laying additional foundation for the subsequent mathematical development of computational mechanics in the later Chapters.

2.1. Reimann and Lebesgue Integrals: Step and Simple Functions

The characteristic function for intervals was described in section 1.2. Similarly, the **characteristic function** of the set S is the function χ_S defined by

$$\chi_S = \begin{cases} 1, \text{ if x is in S} \\ \\ 0, \text{ if x is not in S} \end{cases} \qquad (2.1.1)$$

Let $a_j < b_j$ and $S_j = [a_j, b_j]$. A step function ψ is a finite linear combination of characteristic functions of intervals,

$$\psi = \sum_{j=1}^{n} \lambda_j \chi_{S_j} \qquad (2.1.2)$$

where $\lambda_j \in \mathbf{R}$. Then the Riemann integral of ψ is given by

$$\int \psi = \sum_{j=1}^{n} \lambda_j (b_j - a_j) \qquad (2.1.3)$$

(It is noted in (2.1.3) that the integration does not include a "dx" notation.)

The Lebesgue integral can be obtained analogously except that the characteristic functions are applied to sets S_j which are **measurable** with measure mS_j (where these terms will be defined below). Then for a **simple function**

$$\psi = \sum_{j=1}^{n} \lambda_j X_{S_j} \tag{2.1.4}$$

and the Lebesgue integral of ψ is

$$\int \psi = \sum_{j=1}^{n} \lambda_j \, m \, X_{S_j} \tag{2.1.5}$$

Although the Lebesgue and Riemann integrals often show many similarities, the differences in theory and application are sufficiently significant that much of theory of generalized Fourier series become feasible only upon the development of the Lebesgue integral.

2.2. Lebesgue Measure

From (2.1.4) and (2.1.5) it is seen that the Lebesgue integral utilizes sets S_j which are not necessarily intervals and which require that the measure mS_j exist in some sense.

DEFINITION 2.2.1 (Lebesgue Outer Measure)
Let S be a set. The Lebesgue **outer measure** (or exterior measure) of S, denoted by $m_e(S)$, is the greatest lower bound of the measure of all open sets v such that $S \subset v$.

DEFINITION 2.2.2
The **interior measure** $m_i(S)$ of a set S is the least upper bound of the lengths of all closed sets T such that $T \subset S$.

DEFINITION 2.2.3 (Measurable Set)
Let S be a set. The set S is **measurable** with respect to an outer measure if for any set T

$$m_e(T) = m_e(T \cap S) + m_e(T \cap S^c)$$

The **measure** of S is $mS = m_e(S)$.

If $m_e(S) = m_i(S)$, then the set S is said to be **measurable.** It can be shown that a set S is measurable if and only if for any $\varepsilon > 0$ there is an open set v such that $S \subset v$ and $m_e (v - S) < \varepsilon$; similarly, there is a closed set $T \subset S$ such that $m_e (S - T) < \varepsilon$; finally, there are sets v and T such that $T \subset S \subset v$ and $m_e (v - T) < \varepsilon$.

Several important theorems follow from the definition of measure:

(i)   If a set S is measurable, then S^c is measurable.

(ii)   If S_1 and S_2 are measurable sets, then the union $S_1 \cup S_2$ and the intersection $S_1 \cap S_2$ are measurable.

(iii) If S_1 and S_2 are measurable sets, then $m(S_1 \cup S_2) = m(S_1) + m(S_2) - m(S_1 \cap S_2)$.

(iv)   Let S_1, S_2, $S_3, \cdots$ be a sequence of measurable sets. Then $\bigcup_{n=1}^{\infty} S_n$ is measurable. If the S_n are mutually disjoint, $m(\bigcup_{n=1}^{\infty} S_n) = \sum_{n=1}^{\infty} m(S_n)$. Furthermore, for S_n not mutually disjoint, $m(\bigcup_{n=1}^{\infty} S_n) \le \sum_{n=1}^{\infty} m(S_n)$.

(v)   Any open set S in **R** can be written in terms of a countable union of disjoint open intervals S_i called **component intervals.** Because each open interval $S_i = (a_i, b_i)$ has length $(b_i - a_i)$, the measure of any open set in **R** is the length of S; that is, $m(S) = \sum_{i=1}^{\infty} m(S_i)$.

DEFINITION 2.2.4 (Measure Zero)

If a set S has $m_e(S) = 0$, then $m(S) = 0$; that is, S has **measure zero.**

DEFINITION 2.2.5 (Almost Everywhere, ae)

A property or function which applies everywhere on a set S except for a subset $S^* \subset S$ such that $m(S^*) = 0$ is said to apply **almost everywhere** or **ae.** (Considerations of ae apply in computational analysis of cracks, nonhomogeneity barriers, and other areas).

Example 2.2.1

Let S be a countable set of real numbers $S = \{S_1, S_2, S_3, \cdots\}$. Then each $S_i \in S$ can be enclosed by an interval I_i of length $\varepsilon/2^i$, for $\varepsilon > 0$, with $S_i \in I_i$. Then the measure of $T = \bigcup_{i=1}^{\infty} I_i$ is

$$mT \leq \sum_{i=1}^{\infty} mI_i = \varepsilon \sum_{i=1}^{\infty} 2^i = \varepsilon$$

Letting $\varepsilon \to 0$ implies $0 \leq mS \leq \varepsilon$ and $mS = 0$.

Example 2.2.2

Let S be a measurable set with complement S^c. Show S^c is measurable:

For any set T, $T \cap S$ and $T \cap S^c$ are disjoint sets and
$$m_e(T) = m_e(T \cap S) + m_e(T \cap S^c).$$
However $m_e(T) = m_e(T \cap S^c) + m_e(T \cap (S^c)^c)$
$$= m_e(T \cap S^c) + m_e(T \cap S) = m_e(T)$$
Thus, S^c is a measurable set.

Example 2.2.3

Suppose $m_e(S) = 0$ for some set S. Show S is measurable:

Let T be any set. Then $m_e(T \cap S) \geq 0$ and $m_e(T \cap S^c) \geq 0$. Because $m_e S = 0$, then $(T \cap S) \subset S$ implies $m_e(T \cap S) = 0$.

Thus, $T \cap S^c \subset T$ implies
$$m_e(T) \geq m_e(T \cap S^c) = m_e(T \cap S) + m_e(T \cap S^c)$$
Therefore, S is measurable, and $mS = 0$.

2.3. Measurable Functions

Not only must the set S (over which the integration is to occur) have special characteristics (be measurable), the functions which are to be integrated over S must be qualified as well.

DEFINITION 2.3.1 (Measurable Function)

Let S be a measurable set and the real function $\phi(x)$ be defined on S. Then ϕ is said to be a **measurable function** (or measurable) on S if for any $\alpha \epsilon \mathbf{R}$ the set $\{x \epsilon S: \phi(x) > \alpha\}$ is measurable.

Several variations of Definition 2.3.1 are possible. The function is measurable on S if for any $\alpha \epsilon \mathbf{R}$ the sets $\{x \epsilon S: \phi(x) \geq \alpha\}$, $\{x \epsilon S: \phi(x) \leq \alpha\}$, $\{x \epsilon S: \phi(x) < \alpha\}$ are measurable. Or for $\alpha < \beta$ and $\alpha, \beta \epsilon \mathbf{R}$, ϕ is measurable if the sets $\{x \epsilon S: \alpha < \phi(x) < \beta\}$, $\{x \epsilon S: \alpha \leq \phi(x) \leq \beta\}$, and so forth, are measurable.

Several theorems supply useful information regarding a function ϕ that is measurable on the measurable set S:

(i)   If $S^* \subset S$, then ϕ is measurable on the S^*.

(ii)   A constant function ϕ is measurable for $\lambda \epsilon \mathbf{R}$.

(iii) $(\lambda\phi)$, $(\phi + \lambda)$, and ϕ^2 are measurable on S.

(iv)   Let ϕ_1 and ϕ_2 be measurable on S, then $(\phi_1 + \phi_2)$, $(\phi_1 - \phi_2)$, $(\phi_1 \phi_2)$ and (ϕ_1/ϕ_2) are measurable on S where for ϕ_1/ϕ_2 it is assumed $\phi_2 \neq 0$.

(v)   If ϕ_1 is measurable on S and $\phi_2 = \phi_1$ ae, then ϕ_2 is measurable on S.

(vi)   A continuous function is measurable.

(vii) A function which is continuous over S except for a countable number of discontinuities, is measurable.

Fortunately, the class of functions normally dealt with in computational mechanics applications and the sets involved are both measurable. However in studying whether the type of approximation function applied to the operator equation has a possibility to converge to the exact solution, the reliance upon integration theory becomes important in proving convergence of generalized Fourier series.

2.4. The Lebesgue Integral

The Lebesgue integral is similar in concept to the Riemann integral in that a partition is utilized in the integration. In the Riemann integral, the domain is partitioned; in the Lebesgue integral, the range of the integrable function ϕ is partitioned.

If ϕ is a function that is Riemann integrable on S = [a,b], then it is also Lebesgue integrable and the two integrals equal each other. However, a Lebesgue integrable function is not necessarily Riemann integrable.

2.4.1. Bounded Functions

Let $\alpha < \phi(x) < \beta$ be defined on [a,b], such that α and β are in **R**. Partition (α,β) into n intervals by points ξ_j such that $\alpha = \xi_0 < \xi_1 < \cdots < \xi_{n-1} < \xi_n = \beta$. Define the sets S_j in the domain by $S_j = \{x \in [a,b]: \xi_{j-1} \leq \phi(x) < \xi_j\}$ for $j = 1,2,\cdots,n-1$, and $S_n = \{x \in [a,b]: \xi_{n-1} \leq \phi(x) \leq \xi_n\}$. Then the **upper sum** is

$$U = \sum_{j=1}^{n} \xi_j\, m(S_j)$$

and the **lower sum** is

$$L = \sum_{j=1}^{n} \xi_{j-1}\, m(S_j)$$

Let $\int_a^{\overline{b}} \phi(x)dx$ be the greatest lower bound of all possible values of U for all possible partitions, and let $\int_{\underline{a}}^b \phi(x)dx$ be the least upper bound of all possible values of L for all possible partitions. If the above two integrals are equal, then $\phi(x)$ is said to be **Lebesgue integrable** on [a,b] and the integral is noted by $\int_a^b \phi(x)dx$ or simply $\int_a^b \phi$.

Several useful theorems follow:

(i)   Let S be a measurable set in [a,b]. Let ϕ be bounded and measurable on S. Then the Lebesgue integral exists and is noted by $\int_S \phi$. Also, $\int_S |\phi| < \infty$.

(ii) If mS = 0, then $\int_S \phi = 0$.

(iii) If $S = S_1 \cup S_2$ where all the sets are measurable and $S_1 \cap S_2 = \emptyset$, then $\int_S \phi = \int_{S_1} \phi + \int_{S_2} \phi$.

(iv) Let S be a measurable set and ϕ be measurable on S.

Then

· $\int_S \lambda = \lambda m(S)$ where λ a constant in **R**

· $\int_S \lambda\phi = \lambda \int \phi$

· If $\alpha \le \phi \le \beta$, then $\alpha m(S) \le \int \phi \le \beta m(S)$

(v) Let ϕ_1 and ϕ_2 be bounded and measurable on a measurable set S. Then

· $\left| \int_S \phi_1 \right| \le \int_S \left| \phi_1 \right|$

· If $\phi_1 \le \phi_2$, then $\int_S \phi_1 \le \int_S \phi_2$

· $\int_S (\phi_1 + \phi_2) = \int_S \phi_1 + \int_S \phi_2$

· If $\phi_1 = \phi_2$ ae, then $\int_S \phi_1 = \int_S \phi_2$

· $\int_S \phi_1 = 0$ and $\phi_1 \ge 0$ ae on S, then $\phi_1 = 0$ ae on S

(vi) Let ϕ be bounded and measurable on $S = \bigcup_{j=1}^{\infty} S_j$ where the S_j are all measurable sets and are mutually disjoint.
Then $\int_S \phi = \sum_{j=1}^{\infty} \int_{S_j} \phi$

(vii) Let ϕ_1 and ϕ_2 be bounded and measurable on $S = (a,b)$ and

$\int_S (\phi_1 - \phi_2)^2 = 0$. Then $\phi_1 = \phi_2$ ae in S.

(viii) Let S be a measurable set and let the sequence of functions $<\phi_j>$, $j = 1,2,\cdots,n$ be bounded and measurable on S and λ_j, $j = 1,2,\cdots,n$ be constants in **R**.
Then $\int_S \sum_{j=1}^{n} \lambda_j \phi_j = \sum_{j=1}^{n} \lambda_j \int_S \phi_j$

2.4.2. Unbounded Functions

Section 2.4.1 addressed bounded functions and Lebesgue integration. In this section, unbounded functions are considered. Such complications are involved in numerically modeling singularities in the problem domain or boundary. Let S be a measurable set and let ϕ be divided into positive and negative parts by

$$\phi^+(x) = \begin{cases} \phi(x), \ x \in S \text{ such that } \phi(x) \geq 0 \\ \\ 0, \quad \text{otherwise} \end{cases}$$

$$\phi^-(x) = \begin{cases} \phi(x), \ x \in S \text{ such that } \phi(x) < 0 \\ \\ 0, \quad \text{otherwise} \end{cases}$$

Then ϕ^+ and ϕ^- are both nonnegative functions such that $\phi = \phi^+ - \phi^-$. Consider first ϕ^+. Define functions f_p, for each $p > 0$, $p \in \mathbf{R}$, by

$$f_p(x) = \begin{cases} \phi(x), \ x \in S \text{ such that } \phi(x) \leq p \\ \\ p, \quad \text{otherwise} \end{cases}$$

Then f_p is bounded, measurable, and Lebesgue integrable. The Lebesgue integral $\int_S \phi^+$ is defined by

$$\int_S \phi^+ = \lim_{p \to \infty} \int_S f_p$$

If the limit is finite, $\int_S \phi^+$ is said to exist or ϕ^+ is integrable on S. If $\int_S \phi^+$ is infinite, ϕ^+ is said to be **not integrable**.

The Lebesgue integral of $\int_S \phi^-$ follows analogously. Then,

$$\int_S \phi = \int_S \phi^+ - \int_S \phi^-.$$

Note that $\int_S \phi$ is well defined in that infinity is not possible (by definition) for either integral on the right hand side of the equality.

Let ϕ_1 and ϕ_2 be measurable on the measurable set S where ϕ_1 or ϕ_2 are not necessarily bounded. The following theorems are useful:

(i) Let $|\phi_1(x)| \leq \phi_2(x)$ ae, where $x \in S$ and suppose ϕ_2 is integrable on S. Then ϕ_1 is integrable on S and

$$\int_S |\phi_1| \leq \int_S \phi_2.$$

(ii) ϕ_1 is integrable on S if and only if $|\phi_1|$ is integrable on S. We have $|\int_S \phi_1| \leq \int_S |\phi_1|$ if either condition applies.

(iii) Let $\int_S \phi_1$ exist (i.e., $\int_S \phi_1$ is finite). Then ϕ_1 is finite ae on S.

(iv) Let S^* be a measurable subset of S. Then if $\int_S \phi_1$ exists, (i.e., is finite), then

$$\int_{S^*} \phi_1 \text{ exists and } \int_{S^*} |\phi_1| < \int_S |\phi_1|$$

(v) Theorems (ii), (iii), (iv), (v) of Section 2.4.1 (regarding bounded functions) apply to unbounded functions.

Example 2.4.1.

A set S in **R** is said to be **denumerable** if each element in S can be put into a one-to-one correspondence with the positive integers. A set which is either empty (has no element), finite, or denumerable is said to be **countable**. The positive rational numbers Q^+ where $Q^+ = \{x \in \mathbf{R} : x = \frac{m}{n}$ where m and n are positive integers and $n \neq 0\}$ are countable. To show this, arrange all the elements in Q^+ as follows:

$$Q^+ = \{0; 1; \frac{1}{2}; \frac{1}{3}; \frac{2}{3}; \frac{1}{4}, - , \frac{3}{4}, - ; \frac{1}{5}, \frac{2}{5}, \frac{3}{5}, \frac{4}{5}, - ; ...\}$$

In the above set, the rationals are arranged according to common denominators. It is noted that the blanks indicate rationals already accounted for; for example, $\frac{2}{4} = \frac{1}{2}$ and $\frac{4}{4} = \frac{1}{1} = 1$. This arrangement of Q^+ enables the positive rationals to be counted. Indeed, a computer program could be prepared to sweep through the above set, and any rational you choose would eventually be printed out. Hence, Q^+ is countable.

Example 2.4.2

Consider the unit interval $[0,1]$. Construct a function ϕ on $[0,1]$ as follows: Take a piece of paper of height 1 and width 1/2. Cut the paper into halves, always preserving the height with a length of 1. Center one of two pieces of paper in the interval $[0,1]$ so that the subinterval $[3/8, 5/8]$ is covered. Now cut the remaining piece of paper into quarters, resulting in 4 pieces of paper of height 1 and width 1/16. Place two of these pieces in the center of the two intervals $[0, 3/8)$ and $(5/8, 1]$. Continuing this procedure, the interval $[0,1]$ will have the function ϕ defined such that any subinterval S^* (of any length) contained in $[0,1]$ will have function values of $\phi(x) = 1$ for some $x \in S^*$. In fact, most intervals $S^* \subset [0,1]$ will have function values of $\phi(x_1) = 0$ and $\phi(x_2) = 1$ for some x_1 and $x_2 \in S^*$.

The Riemann integral applied to $\phi(x)$ on $[0,1]$ will result in $\int_0^1 \phi(x)dx$ being undefined since the upper Riemann sum Σ^* is given by

$$\Sigma^* = \sum_{j=1}^{n} (1) (\Delta x_j), \text{ for Riemann intervals } \Delta x_j.$$

and the lower Riemann sum Σ_* contains several intervals Δx_j where $\min(\phi(x)) = 0$. Thus, $\Sigma^* \neq \Sigma_*$ and the Riemann integral is undefined.

The Lebesgue intergral, on the other hand, solves this integration problem by noting for $S = [0,1]$

$$\int_S \phi = (0)m \{x \in S: \phi(x) = 0\} + (1)m \{x \in S: \phi(x) = 1\}$$

$$= m \{x \in S: \phi(x) = 1\}$$

$$= 1/2.$$

Example 2.4.3

Let ϕ be a nonnegative measurable function defined on the measurable set S. Then there exists a sequence $<\phi_n>$ of functions measurable on S such that (i) the $<\phi_n>$ are monotonically increasing (i.e., $\phi_{n+1}(x) \geq \phi_n(x)$ for $x \in S$ and all n); (ii) $\phi_n(x) \geq 0$; (iii) as $n \to \infty$, $\phi_n(x) \to \phi(x)$; (iv) each ϕ_n has only a finite number of values; (v) $\phi_n(x) \in \mathbf{R}$.

The $<\phi_n>$ are constructed by defining sets (we let $<\phi_n>$ denote $<\phi_n; n=1,2,\cdots>$), $S_{jn} = \{x \in S: j/2^n \le \phi(x) < (j+1)/2^n\}$ and for $j = n2^n$, $S_{jn} = \{x \in S: \phi(x) \ge n\}$. These sets have the properties that the sets S_{jn} are all mutually disjoint and yet the union of the S_{jn} equals the parent set, S. Letting $\phi_n = j/2^n$ on S_{jn} results in the desired sequence of approximation functions.

From Example 2.2.1 and Example 2.4.3, the set of rationals has measure zero. Thus, on any interval in R, it is the irrationals that has the measure equal to the length of the interval. Thus, the irrationals are uncountable from Example 2.2.1. It can be shown that unmeasurable sets exist by using sets constructed as a countably infinite union of disjoint sets of irrational numbers where each disjoint set has the same measure and, additionally, is itself a subset of a finite parent interval.

2.5. Key Theorems in Integration Theory

In using any numerical method, the engineer develops a sequence of approximation functions $<\phi_n>$ by repeated tries at solving the governing operator equation. For example, ϕ_{n+1} may be nearly the same approximation as ϕ_n except that for ϕ_{n+1}, additional basis functions, interpolation points, or collocation points were added to the computation.

Generally, the analyst evaluates his work effort by observing how much the computational results change with the use of additional computational effort in the model; and if the differences between the two modeling efforts are small, then it is usually concluded that a good approximation has been achieved.

Mathematically, the above "convergence" criteria is stated as $||\phi_n - \phi_{n+1}||$ is small where typically the norm is the judgement of the analyst, and $||\phi_n - \phi_{n+1}||$ is evaluated at a few discrete points in the problem domain.

In this section, a few key theorems are presented which will be of importance in establishing convergence properties of a sequence of functions.

Theorem 2.5.1

Let $<\phi_n>$ be a sequence of nonnegative, monotonically increasing functions measurable on the measurable set S. Suppose that $\phi_n(x) \to \phi(x)$, as $n \to \infty$. Then as $n \to \infty$, $\int_S \phi_n \to \int_S \phi$.

Example 2.5.1

Let $<\phi_n>$ be a sequence defined on $[0,1]$ by

$$\phi_n(x) = \begin{cases} 1, & \text{for } 0 < x \le 1 - \dfrac{1}{n} \\ \\ 0, & \text{otherwise} \end{cases}$$

Let $\phi(x)$ be defined by $\phi(x) = 1$ for $S = \{x: 0 \le x \le 1\}$. Then $<\phi_n>$ is a monotonically increasing sequence of measurable functions and as $n \to \infty$, $\phi_n(x) \to \phi(x)$ for $x \in [0,1]$. Then $\int_S \phi_n = (1 - \frac{1}{n})$. Then as $n \to \infty$, $\int_S \phi_n \to 1 = \int_S \phi$.

Example 2.5.2

Define a sequence $<\phi_n>$ on $[0,\infty)$ by

$$\phi_n(x) = \begin{cases} \dfrac{1}{n}, & \text{for } 0 \le x \le n \\ \\ 0, & \text{otherwise} \end{cases}$$

and define $\phi(x) = 0$ for $x \ge 0$. Then as $n \to \infty$, $\phi_n(x) \to \phi(x)$. But $\int \phi_n = \frac{1}{n}(n) = 1$ and $\int \phi = 0$. Hence as $n \to \infty$, $\int \phi_n \ne \int \phi$, showing that the Monotone Convergence theorem may only apply to a monotonically increasing sequence of measurable functions.

Theorem 2.5.2

Let $<\phi_n>$ be a sequence of functions measurable on the measurable set S such that as $n \to \infty$, $\phi_n(x) \to \phi(x)$ ae. If there is a non-negative function f which is measurable on S such that $|\phi_n(x)| \le f(x)$ for $n = 1,2,...$, and for all x in S, then for $n \to \infty$, $\int_S \phi_n = \int_S \phi$.

Theorem 2.5.3 (Egorov's Theorem)

Let $<\phi_n>$ be a sequence of functions measurable on a set S such that as $n\to\infty$, $\phi_n(x)\to\phi(x)$ ae on S where $\phi(x)$ is bounded. Then for any real number $\delta>0$ there exists a set S^* in S such that $mS^* > mS - \delta$ and as $n\to\infty$, $\phi_n(x)\to\phi(x)$ uniformly.

2.6. L_p Spaces

The set of all functions ϕ defined on S (where $S \subset \mathbf{R}$) such that the function $|\phi(x)|^p$ for $p > 1$ is Lebesgue integrable is denoted by $L_p(S)$. That is, a function ϕ is an element of $L_p(S)$ if and only if

$$\int_S |\phi|^p < \infty.$$

Example 2.5.3

Let $p = 2$ in the above description of L_p spaces. Then a function is in $L_2(S)$ if and only if $\int_S \phi^2 < \infty$.

2.6.1. m-Equivalent Functions

Let ϕ_1 and ϕ_2 be functions measurable on the measurable set S. Then if $\phi_1 = \phi_2$ ae on S, ϕ_1 and ϕ_2 are said to be **m-equivalent**. That is, $m\{x \in S: \phi_1(x) \neq \phi_2(x)\} = 0$. Because many functions can be m-equivalent, the notation $[\phi]$ is used to designate the equivalence class of functions which are m-equivalent to ϕ on S.

The Lebesgue space $L_1(S)$ with the usual Lebesgue measure (m) consists of elements which are equivalence classes. In $L_1(S)$, the norm is defined in $[\phi]$ by $||[\phi]||_1 = \int_S |\phi|$. Because integration on sets of measure zero is zero, the equivalence class notation $[\phi]$ can be dropped in all future work. However, the reader must recall that all measurable functions being studied are but single elements of an equivalence class of measurable functions.

2.6.2. The Space L_p

Let $1 \leq p < \infty$. Then the space $L_p(S)$ consists of all equivalence classes of functions measurable on the measurable set S. The norm is defined for $\phi \in L_p(S)$ by $||\phi||_p = [\int |\phi|^p]^{\frac{1}{p}}$. Then L_p is a normed linear space whose norm definition guarantees that every Cauchy sequence $< \phi_n >$ of elements $\phi_n \in L_p(S)$ converges to a limit $\phi \in L_p(S)$. Thus $L_p(S)$ is a Banach space.

The following theorems are extensively used in determining bounds on approximation error for various types of numerical techniques.

Theorem 2.5.4 (Holder's Inequality)

Let $\phi_p \in L_p(S)$ and $\phi_q \in L_q(S)$ where $1/p + 1/q = 1$ and $p > 1$. Then (i) the product $\phi_p \phi_q \in L_1(S)$, and (ii) $||\phi_p \phi_q||_1 \leq ||\phi_p||_p ||\phi_q||_q$; that is, $|\int_S \phi_p \phi_q| \leq |\int_S \phi_p|^p|^{1/p} |\int_S |\phi_q|^q|^{1/q}$.

Theorem 2.5.5 (Cauchy-Buynakovskii-Schwarz Inequality)

Let $p = 2$ in Holder's Inequality. Then $q = 2$ and

$$|\int_S \phi_1 \phi_2| \leq \int_S |\phi_1 \phi_2| < ||\phi_1||_2 ||\phi_2||_2$$

where ϕ_1 and ϕ_2 are in $L_2(S)$.

Theorem 2.5.6 (Minkowski's Inequality)

Let ϕ_1 and ϕ_2 be in $L_p(S)$ for $p \geq 1$. Then $\phi_1 + \phi_2$ are in $L_p(S)$ and

$$||\phi_1 + \phi_2||_p \leq ||\phi_1||_p + ||\phi_2||_p$$

Theorem 2.5.7 (Triangle Inequality)

Let $p = 2$ in the Minkowski Inequality. Then
$$||\phi_1 + \phi_2||_2 < ||\phi_1||_2 + ||\phi_2||_2$$

That is,

$$\int_S (\phi_1 + \phi_2)^2 \leq \int_S \phi_1^2 + \int_S \phi_2^2$$

With the previous theorems and definitions, the integration theory that is embodied in the Riemann integral has been generalized for use with sequences of approximations and convergence of these sequences to the problem solution.

2.7. The Metric Space, L_p

Given a space $L_p(S)$, the distance between two vectors (points, or elements) in $L_p(S)$ is defined by

$$D(\phi_1, \phi_2) = ||\phi_1 - \phi_2||_p = [\int_S |\phi_1 - \phi_2| \, p]^{1/p}$$

for ϕ_1 and ϕ_2 in $L_p(S)$. Then it is seen that (i) $D(\phi_1,\phi_2) \geq 0$; (ii) $D(\phi_1,\phi_2) = 0$ implies $\phi_1 = \phi_2$ ae on S (i.e., ϕ_1 and ϕ_2 are in the same equivalence class); (iii) $D(\phi_1,\phi_2) = D(\phi_2,\phi_1)$; and (iv) from Minkowski's Inequality $D(\phi_1,\phi_2) \leq D(\phi_1,\phi_3) + D(\phi_3,\phi_2)$ where ϕ_3 is also in $L_p(S)$.

2.8. Convergence of Sequences

We are now prepared to discuss how well our approximation function ϕ_n approximates the true solution ϕ of an operator equation. As we develop more and more "accurate" approximations, we are generating a sequence of approximations $<\phi_n>$ which, as $n \to \infty$, we hope that $\phi_n \to \phi$.

The first question to be asked is: what is the character of the working space, and then the norm needs to be defined. Because there are several modes of convergence, the type of convergence to be studied needs to be determined.

2.8.1. Common Modes of Convergence

A sequence $<\phi_n> \to \phi$ **uniformly** in **R** if for every $\varepsilon > 0$ there exists a positive integer $N(\varepsilon)$ such that $n \geq N(\varepsilon)$ implies $|\phi_n(x) - \phi(x)| < \varepsilon$ for all $x \in S$. This type of convergence is denoted by **U**.

A sequence $<\phi_n> \to \phi$ **pointwise** if for every $\varepsilon > 0$ and $x \in S$ there exists an $N(\varepsilon,x)$ such that $n \geq N(\varepsilon,x)$ implies $|\phi_n(x) - \phi(x)| < \varepsilon$. This type of convergence is denoted by **P**.

52

Similar to pointwise convergence, **ae convergence** indicates pointwise convergence except for a set $S^* \subset S$ where $m(S^*) = 0$. This convergence is denoted by **AE**.

From the above definitions, $\mathbf{U} \to \mathbf{P} \to \mathbf{AE}$, but no other conclusions can be drawn.

2.8.2. Convergence in L_p

$<\phi_n> \to \phi$ in L_p on set S if for every $\varepsilon > 0$ there exists a $N(\varepsilon)$ such that if m and n $\geq N(\varepsilon)$, then

$$||\phi_m - \phi_n||_p = \left[\int_S |\phi_m - \phi_n|^p \right]^{1/p} < \varepsilon$$

2.8.3. Convergence in Measure (M)

A sequence of measurable functions $<\phi_n> \to \phi$ **in measure** on measurable set S if as $n \to \infty$,

$$m\{x \in S: |\phi_n(x) - \phi(x)| \geq \lambda\} = 0$$

for any $\lambda \geq 0$, $\lambda \in \mathbf{R}$

2.8.4. Almost Uniform Convergence (AU)

Let $<\phi_n>$ and ϕ be functions measurable on the measurable set S. Then $<\phi_n> \to \phi$ **AU** if for any $\delta > 0$ there exists a set $S_\delta \subset S$ such that $<\phi_n> \to \phi$ **U** on the set $S^* = S - S_\delta$, and $m(S_\delta) < \delta$.

2.8.5. Is the Approximation Converging?

Each type of convergence may or may not imply another type of convergence. A convenient summary of the cross-implications in the types of convergence can be made in diagram form as follows:

For any measurable set S,

(3 implications)

If mS is finite, two more implications can be concluded,

(5 implications)

Notice in this case that **AE** only implies **AU** and **M** and not L_p. Also, L_p only implies **M** convergence. Should $<\phi_n>$ be a sequence of functions such that $|\phi_n(x)| \leq f(x)$ for all $x \in S$ and $f(x)$ is integrable on S, then three more implications are added,

(8 implications)

Here, **AU**, **AE** and **M** convergence implies L_p convergence, yet L_p convergence only implies convergence in **M.** This should be remembered by the analyst as many numerical methods solve for the best approximation in the L_p sense (where p = 2).

2.8.6. Counterexamples

It is important to understand the limitations between the various types of convergence. To demonstrate the failures of one type of convergence implying another type of convergence, several classic counterexamples are provided in the following.

Example 2.8.1

$$\text{Let } \phi_n(x) = \begin{cases} n^{-1/p}, & \text{for } 0 \leq x \leq n \\ \\ 0, & \text{for } x \geq 0 \end{cases}$$

Then as $n \to \infty$, $\phi_n(x) \to \phi(x)$ uniformly (or **U**) where $\phi(x) = 0$ for $x \geq 0$. But

$$||\phi_n - \phi||_p = (\int_0^\infty |\phi_n - \phi|^p)^{1/p} = (\int_0^n (\frac{1}{n^p})^p)^{\frac{1}{p}} = (\int_0^n 1/n)^{1/p};$$

and as $n \to \infty$, $||\phi_n - \phi||_p \to 1$. Therefore, **U** does not imply L_p.

Example 2.8.2

$$\text{Let } \phi_n(x) = \begin{cases} n^2, & \text{for } 1/n \leq x \leq 2/n \\ \\ 0, & \text{for } x \geq 0 \end{cases}$$

Then as $n \to \infty$, $\phi_n(x) \to \phi(x) = 0$ pointwise, (P), for $x = 0$. But

$$||\phi_n - \phi||_p = (\int_{1/n}^{2/n} (n^2)^p)^{1/p} = (\int_{1/n}^{2/n} (n^2 p)^{1/p} = (n^2 p - 1)^{1/p} = n^{2-1/p}.$$

Thus as $n \to \infty$, $||\phi_n - \phi||_p \to \infty$. Therefore, P does not imply L_p.

Example 2.8.3

Let $\phi_n(x)$ be the triangle such that $\phi_n(x = 0) = 0$,

$\phi_n(x = \frac{1}{n}) = n^2$, $\phi_n(x = \frac{2}{n}) = 0$, and $\phi_n(x > 2/n) = 0$, for all n

and $x \geq 0$. Let $\phi(x) = 0$ for $x \geq 0$. Then as $n \to \infty$, $\phi_n(x) \to \phi(x)$ pointwise (P). But $||\phi_n - \phi||_1 = (\int_0^{2/n} \phi_n) = n$. Thus as $n \to \infty$, $||\phi_n - \phi||_1 \to \infty$ showing that P does not imply L_p.

Example 2.8.4

Let $\phi_n(x)$ be defined on $S = \{x: x \geq 0\}$ by

$$\phi_n(x) = \begin{cases} 0, \text{ for } 0 \leq x \leq n \\ \text{triangle, where } \phi_n(x = n) = 0, \\ \quad \phi_n(x = n+1) = 1, \phi_n(x = n+2) = 0 \\ 0, \text{ for } x \geq n+2 \end{cases}$$

Let $\phi(x) = 0$ for $x \geq 0$. Then for $\alpha > 0$ and any n, $m\{x \in S: |\phi_n(x) - \phi(x)| \geq \alpha\} = m\{x \in S: \phi_n(x) \geq \alpha\} = 2$; that is, the length of the interval $(n, n+2)$. Thus as $n \to \infty$, $m\{x \in S: |\phi_n(x) - \phi(x) \geq \alpha\} = 2$. Therefore, $\phi_n(x) \to \phi(x)$ pointwise but $\phi_n(x)$ does not $\to \phi(x) = 0$ in measure. Hence, P does not imply M.

Example 2.8.5

From __Example 2.8.2__, ϕ_n does not converge in L_p to ϕ on $S = \{x: x \geq 0\}$. But $m\{x \in S: |\phi_n(x) - \phi(x)| \geq \alpha\} \to 1/n$ as $\alpha \to 0$. And as $n \to \infty$, $m\{x \in S: |\phi_n(x) - \phi(x) \geq \alpha\} \to 0$. Thus, $\phi_n(x) \to \phi(x)$ in measure (**M**), but not in L_p.

Example 2.8.6

Construct the sequence $<\phi_n>$ such that each ϕ_n is a rectangle of height 1 and having decreasing base according to the sequence 1, 1/2, 1/2, 1/3, 1/3, 1/3, $\cdots$. Each rectangle is placed on $S = [0,1]$ so that ϕ_1 covers $[0,1]$, ϕ_2 covers $[0, 1/2]$, ϕ_3 covers $[1/2, 1]$, ϕ_4 covers $[0, 1/3]$, ϕ_5 covers $[1/3, 2/3]$, ϕ_6 covers $[2/3, 1]$, and so forth. Let $\phi(x) = 0$ on S. As $n \to \infty$, the $m\{x \in S: |\phi_n(x) - \phi(x)| > \alpha\} \to 0$. Also, as $n \to \infty$, $||\phi_n - \phi||_p = (\int_S \phi_n^p)^{1/p} \to 0$. But for any $x \in S$, consider the sequence of values $<\phi_n(x)>$ which is composed of zeroes and ones. That is, as $n \to \infty$, each "cycle" of ϕ_n as the rectangle bases shift from covering $x = 0$ to finally covering the point $x = 1$, will cover any point $x \in S$. Hence $\phi_n(x)$ does not pointwise converge anywhere. Thus, $\phi_n \to \phi$ in both **M** and L_p, and yet ϕ_n does not converge to ϕ pointwise, even on a finite measure space.

2.9. Capsulation

This chapter reviewed the basic theory of Lebesgue integration needed to develop the generalized Fourier series theory, and developed the theory fundamental to converging sequences. Because the numerical modeler typically is developing a sequence of approximations $<\phi_n>$ to ϕ on S, he needs to know that $\phi_n \to \phi$ as $n \to \infty$ and if so, by what standard of convergence.

Many numerical methods deal with L_2 convergence and, from section 2.8, it is seen that L_2 convergence only guarantees convergence in measure. But in engineering problems in general, additional hypothesis are available such as continuity, piecewise continuity, and so forth. As a result of these additional hypothesis, additional implications are available other than those shown in the logic diagrams of section 2.8.5. These additional hypothesis will be discussed in Chapter 3 along with the Hilbert space environment, generalized Fourier series, and finite dimensional vector space representations of piecewise continuous functions defined over the problem domain.

CHAPTER 3
HILBERT SPACE AND GENERALIZED FOURIER SERIES

3.0. Introduction

The subjects of inner product, Hilbert space, generalized Fourier series, and vector space representations are all implicitly used in many numerical methods. Consequently, theoretical principles already established can be directly applied to many topics in computational mechanics.

To demonstrate the interplay between these concepts, let the underlying Banach space be $\mathbf{R}^3$ where each element (or vector) $\xi \in \mathbf{R}^3$ is of the form $\xi = (x,y,z)$. Let ξ_1 and ξ_2 be two distinct elements and $\overrightarrow{0\xi_1}$ be the straight line through points $(0,0,0)$ and $\xi_1 = (x_1, y_1, z_1)$. The task is to find a point $\xi^* = (x^*, y^*, z^*)$ on $\overrightarrow{0\xi_1}$, which is closest to the point $\xi_2 = (x_2, y_2, z_2)$. Closest is defined to mean the minimum value of the norm generally used in $\mathbf{R}^3$. The norm used is the ℓ_2 norm (Euclidean norm)

$||\xi|| = ||(x,y,z) - (0,0,0)||_2 = [x^2 + y^2 + z^2]^{1/2}$. Any point ξ on $\overrightarrow{0\xi_1}$, has a distance from ξ_2 calculated from

$D^2(\xi, \xi_2) = ||\xi - \xi_2||^2 = (x - x_2)^2 + (y - y_2)^2 + (x - z_2)^2$. But ξ on $\overrightarrow{0\xi_1}$, can be written as $\xi = \lambda\xi_1$ where $\lambda \in \mathbf{R}$. Thus

$D^2(\xi, \xi_2) = (\lambda x_1 - x_2)^2 + (\lambda y_1 - y_2)^2 + (\lambda z_1 - z_2)^2$

$= \lambda^2 (x_1^2 + y_1^2 + z_1^2) - 2\lambda(x_1 x_2 + y_1 y_2 + z_1 z_2) + (x_2^2 + y_2^2 + z_2^2)$

Using vector dot product notation,

$\xi_1 \cdot \xi_2 = x_1 x_2 + y_1 y_2 + z_1 z_2$, $D^2(\xi, \xi_2) = \lambda^2 \xi_1 \cdot \xi_1 - 2\lambda \xi_1 \cdot \xi_2 + \xi_2 \cdot \xi_2$.

The above result can be obtained more quickly from the definition of the norm by noting $D^2(\xi, \xi_2) = ||\lambda\xi_1 - \xi_2||^2$

$= (\lambda\xi_1 - \xi_2) \cdot (\lambda\xi_1 - \xi_2) = \lambda^2 \xi_1 \cdot \xi_1 - 2\lambda \xi_1 \cdot \xi_2 + \xi_2 \cdot \xi_2$.

Differentiating the equation with respect to λ gives that value of $\lambda \in \mathbf{R}$ which minimizes $D^2(\xi, \xi_2)$; that is $\lambda = (\xi_1 \cdot \xi_2)/(\xi_1 \cdot \xi_1)$. Thus, the point ξ^* on $\overrightarrow{0\xi_1}$, must be given by $\xi^* = \xi_1(\xi_1 \cdot \xi_2)/(\xi_1 \cdot \xi_1)$.

Because ξ^* is closest to the point ξ_2, the line $\overrightarrow{\xi^*\xi_2}$ must be geometrically orthogonal to the line $\overrightarrow{0\xi_2}$. This is described by the dot product $\overrightarrow{\xi^*\xi_2}\cdot\overrightarrow{0\xi_1} = 0$. That is, for $\overrightarrow{\xi^*\xi_2} = (\xi_2 - \lambda\xi_1)$ and $\overrightarrow{0\xi_1} = \xi_1$, $0 = (\xi_2 - \lambda\xi_1)\cdot\xi_1 = \xi_2\cdot\xi_1 - \lambda\xi_1\cdot\xi_1$ giving $\lambda = \xi_2\cdot\xi_1/\xi_1\cdot\xi_1$.

This geometric interpretation of closeness extends immediately to vector spaces $\mathbf{R}^n$ for all $n \geq 1$, and for the ℓ_2 norm, even though there fails to be a visual representation expedient to represent functions on $[a,b]$ as vectors of arbitrary dimension n in order to estimate the generalized Fourier coefficients. For example, let $E = \{x: 0 \leq x \leq 1\}$ and subdivided $E = [0,1]$ with 11 collocation points located at coordinates 0, 1/10, 2/10,$\cdots$, 9/10, 1. Let f be a function defined on $S = [0,1]$. Then the vector representation $\mathbf{F}$ of f in R^n is given by $\mathbf{F} = (f(0),$ $f(1/10),\cdots,f(1))$. In general, as n gets large, $\mathbf{F}$ represents $f(x)$ on S more accurately in a step function sense. Obviously, n may have to be large in order to capture in $\mathbf{F}$ all the important characteristics of $f(x)$.

3.1. Inner Product and Hilbert Space
DEFINITION 3.1.1. (Hilbert Space)

A linear space S on which an inner product is defined is called an **inner product space**. If S is also complete, then S is called a **Hilbert space** and the norm is defined by $\|\xi\|^2 = (\xi,\xi)$ for $\xi \in S$.

Several properties are associated to an inner product space S.

(i) Let ξ_1 and ξ_2 be in S. Then $|(\xi_1,\xi_2)| \leq \|\xi_1\| \|\xi_2\|$.
     Note that the norm follows from Definition 3.1.1.

(ii) Let ξ_1 and ξ_2 be in S. Then the **angle** α between ξ_1 and ξ_2 is given by
$$\alpha = \cos^{-1}(\xi_1,\xi_2)/(\|\xi_1\| \|\xi_2\|)$$

(iii) Elements ξ_1 and ξ_2 in S are **orthogonal** if and only if $(\xi_1,\xi_2) = 0$. That is, $\alpha = \pi/2$.

(iv) Similar to vector space $\mathbf{R}^n$, the orthogonal projection of ξ_1 onto ξ_2 is given by $\xi_1(\xi_1,\xi_2)/(\xi_1,\xi_1)$.

(v) A weighted inner product is an inner product and is defined in $\mathbf{R}^n$ by $(\xi_1,\xi_2) = w_1x_1y_1 + w_2x_2y_2 +\cdots+ w_nx_ny_n$, where the w_i are positive numbers.

DEFINITION 3.1.2. (L_2 Space)

The space of all real valued functions f such that

$$||f||^2 = \int_S f^2 < \infty \text{ is } L_2(S).$$

DEFINITION 3.1.3. (Inner Product in $L_2(S)$)

Let ϕ_1 and ϕ_2 be in $L_2(S)$. The inner product is defined by

$$(\phi_1, \phi_2) = \int_S \phi_1 \phi_2.$$

DEFINITION 3.1.4. (Orthogonal Functions)

Let $S = (a,b)$. Let ϕ_1 and ϕ_2 be in $L_2(S)$. Then ϕ_1 and ϕ_2 are orthogonal in S if $(\phi_1, \phi_2) = \int_a^b \phi_1(x) \phi_2(x)dx = 0$. A set of functions $\phi_1, \phi_2, \cdots$ in $L_2(S)$ is said to be an orthogonal set in S if $(\phi_i, \phi_j) = 0$ for $i \neq j$.

Example 3.1.1

A classic example of orthogonal functions are the functions sinmx for $m = 1,2,\cdots$ on $S = (-\pi, \pi)$. Then $\int_{-\pi}^{\pi} \sin mx \sin nx \, dx = 0$ for all $m \neq n$.

DEFINITION 3.1.5. (Orthonormal Functions)

Let $\{f_n\}$ be an orthogonal set of functions in S such that $||f_i|| \neq 0$ for any i, and define functions $\phi_n(x) = f_n(x)/||f_n||$. Then $\{\phi_n\}$ is said to be an <u>orthonormal set</u> of functions in S where $(\phi_i, \phi_j) = \begin{cases} 1, & i = j \\ 0, & i \neq j \end{cases}$

Example 3.1.2.

The set of functions $\{\sin mx/\sqrt{\pi}; \ m = 1,2,\cdots\}$ is orthonormal in $(-\pi,\pi)$.

Example 3.1.3.

Let $< f_i(k) >$ be a sequence of orthonormal functions on $S = [a,b]$. Let an approximation function be $\hat{\phi}(x) = \sum_{i=1}^{\infty} \lambda_i f_i(x)$. Suppose $\hat{\phi}(x)$ converges uniformly to a function $f(x)$ on S in $L_2(S)$.

From the properties of uniform convergence, the approximation $\hat{\phi}(x)$ is arbitrarily close in $f(x)$ uniformly on S and hence

$$f(x) = \sum_{i=1}^{\infty} \lambda_i f_i(x), \ x \in S.$$

Then by the orthonormal properties of $<f_i(x)>$ in $L_2(S)$,

$$\int_a^b f(x)\, f_j(x)\, dx = \int_a^b \sum_{i=1}^{\infty} \lambda_i f_i(x)\, f_j(x)\, dx$$

$$= \sum_{i=1}^{\infty} \lambda_i \int_a^b f_i(x)\, f_j(x)\, dx$$

$$= \sum_{i=1}^{\infty} \lambda_i (f_i, f_j) = \lambda_j$$

Thus, each $\lambda_j = \int_a^b f(x)\, f_j(x)\, dx = (f_i, f_j)$.

Example 3.1.4.

Let $<f_i>$ be a sequence of orthonormal functions on $S = [a,b]$. Let f be in $L_2(S)$. An approximation function $\hat{\phi}(x)$ is

$$\hat{\phi}(x) = \sum_{i=1}^{n} \lambda_i f_i(x)$$

where

$$\lambda_i = \int_a^b f(x)\, f_i(x)\, dx$$

Then with respect to $L_2(S)$,

$$||f(x) - \hat{\phi}(x)||^2 = \int_a^b (f(x) - \hat{\phi}(x))^2 \, dx$$

$$= \int_a^b f^2(x) \, dx - 2 \sum_{i=1}^n \lambda_i \int_a^b f(x)f_i(x) \, dx$$

$$+ \sum_{i=1}^n \sum_{j=1}^n \lambda_i \lambda_j \int_a^b f_i(x)f_j(x) \, dx$$

$$= \int_a^b f^2(x) \, dx - 2 \sum_{i=1}^n \lambda_i^2 + \sum_{i=1}^n \lambda_i^2$$

$$= (f,f) - \sum_{i=1}^n (f,f_i)$$

Because $||f(x) - \hat{\phi}(x)||^2 \geq 0$,

$$(f,f) \geq \sum_{i=1}^n (f,f_i) = \sum_{i=1}^n \lambda_j^2$$

The previous inequality is independent of n, and

$$(f,f) \geq \sum_{i=1}^\infty (f,f_i)$$

which is **Bessel's Inequality.**

Note that if $\hat{\phi}(x) = \sum_{i=1}^\infty \lambda_i f_i(x)$, and $||f(x) - \hat{\phi}(x)|| = 0$, then $(f,f) = \sum_{i=1}^\infty (f,f_i)$

which is the **Parseval's Identity.**

Should $(f,f) - \sum_{i=1}^{n}(f,f_i) = 0$, then in $L_2(S)$, $||f(x) - \hat{\phi}(x)|| = 0$, and no further improvement with respect to the $L_2(S)$ norm is achieved by further addition of basis functions to be used in the approximation. The above Bessel's inequality is useful in the evaluation of L_2 approximation accuracy.

Example 3.1.5

Let $\{f_i\} = \{1,x\}$ and $\Omega_1 = [0,1]$, $\Omega_2 = [-1,1]$.

In $L_2(\Omega_1)$,
$$(f_1,f_2) = \int_0^1 x\,dx = 1/2$$

In $L_2(\Omega_2)$,
$$(f_1,f_2) = \int_{-1}^1 x\,dx = 0$$

Thus, orthogonality of elements may depend upon the underlying domain.

3.2. Best Approximations in an Inner Product Space

From the previous examples, the best approximation is an issue of both definition (of the inner-product) and problem setting (operator, domain, auxilliary conditions, among other topics). The approximation setting is to establish a family of linearly independent trial elements, or a basis, and to generate a sequence of approximations $<\hat{\phi}_n>$ where n is the number of basis elements or vectors used in $\hat{\phi}_n$. To improve approximation accuracy (in the meaning of the defined norm), n is increased by a judicial choice of additional elements in the basis. For each $\hat{\phi}_n$, in general terms,

$$\hat{\phi}_n = \sum_{i=1}^{n} \lambda_i f_i$$

where the $\{f_i\}$ are the elements in the linearly independent basis that span a linear space S_n of dim $S_n = n$. The best approximation, $\hat{\phi}_n$, of some element ϕ not in S_n is achieved when

$$\|\phi - \hat{\phi}_n\| < \|\phi - s\| \qquad (3.2.1)$$

for all elements $s \in S_n$.

Development of the best approximation in S_n is given by

$$\hat{\phi}_n = \sum_{i=1}^{n} c_i f_i$$

where the real numbers c_i, $i = 1,2,\cdots,n$ solve the system of linear equations

$$\begin{bmatrix} (f_1,f_1) & (f_1,f_2) & \cdots & (f_1,f_n) \\ (f_2,f_1) & (f_2,f_2) & \cdots & (f_2,f_n) \\ \vdots & \vdots & & \vdots \\ (f_n,f_1) & (f_n,f_2) & \cdots & (f_n,f_n) \end{bmatrix}_{n \times n} \begin{bmatrix} c_1 \\ c_2 \\ \vdots \\ c_n \end{bmatrix}_{n \times 1} = \begin{bmatrix} (f_1,\phi) \\ (f_2,\phi) \\ \vdots \\ (f_n,\phi) \end{bmatrix}_{n \times 1} \qquad (3.2.2)$$

Should the $\{f_i;\ i = 2,\cdots,n\}$ be orthonormal elements, then (3.2.2) simplifies to

$$\begin{bmatrix} 1 & 0 & \cdots & 0 \\ 0 & 1 & \cdots & 0 \\ \vdots & \vdots & & \vdots \\ 0 & 0 & \cdots & 1 \end{bmatrix}_{n \times n} \begin{bmatrix} c_1 \\ c_2 \\ \vdots \\ c_n \end{bmatrix}_{n \times 1} = \begin{bmatrix} (f_1,\phi) \\ (f_2,\phi) \\ \vdots \\ (f_n,\phi) \end{bmatrix}_{n \times 1} \qquad (3.2.3)$$

or simply

$$c_i = (f_i,\phi),\ i = 1,2,\cdots,n. \qquad (3.2.4)$$

An arbitrary basis $\{f_i\}$ of the linear space S_n can be orthonormalized with respect to the inner-product $(\,,\,)$ by the **Gram-Schmidt** process. The algorithm proceeds as follows:

$$\hat{g}_1 = f_1$$

$$g_1 = \hat{g}_1 / ||\hat{g}_1||$$

$$\hat{g}_2 = f_2 - (f_2, g_1)g_1$$

$$g_2 = \hat{g}_2 / ||\hat{g}_2|| \tag{3.2.5}$$

$$\hat{g}_3 = f_3 - (f_3, g_1)g_1 - (f_3, g_2)g_2$$

$$g_3 = \hat{g}_3 / ||\hat{g}_3||$$

$$\vdots$$

$$\hat{g}_n = f_n - \sum_{i=1}^{n-1} (f_n, g_i)g_i$$

$$g_n = \hat{g}_n / ||\hat{g}_n||$$

From the above algorithm, the $\{g_i \; ; \; i = 1,2,\cdots,n\}$ forms another basis for S_n due to there being n linearly independent elements, g_i, $i = 1,2,\cdots n$, with each element $g_i \in S_n$.

Thus, any element $s \in S_n$ is a linear combination of the elements in the basis $\{g_i\}$,

$$s = \sum_{i=1}^{n} C_i g_i$$

where for each g_j, $j = 1,2,\cdots,n$

$$(s, g_j) = (\sum_{i=1}^{n} C_i g_i, g_j)$$

$$= \sum_{i=1}^{n} (C_i g_i, g_j) \tag{3.2.6}$$

$$= \sum_{i=1}^{n} C_i (g_i, g_j)$$

$$= C_j$$

Thus for $s \in S_n$,

$$s = \sum_{i=1}^{n} (s, g_i) \, g_i \tag{3.2.7}$$

And from (3.2.4) and (3.2.5), for $\phi \notin S_n$, the best approxiation of ϕ for any $\hat{\phi}_n \in S_n$ is

$$\hat{\phi}_n = \sum_{i=1}^{n} (\phi, g_i) \, g_i \tag{3.2.8}$$

As additional basis elements are added, dim S_n increases. Generally, the basis elements of S_n are chosen to also be elements of some parent space S of which S_n is a subspace. Then, as dim S_n increases, the approximation of $\phi \in S$ by $\hat{\phi}_n \in S_n$ improves the accuracy. From Example 3.1.4, Bessel's inequality provides that approximation error does not increase with increasing dim S_n. And from (3.2.7), should $\phi \in S_n$, then there is no approximation error with respect to (,).

DEFINITION 3.2.1. (Generalized Fourier Series)

Let $\{g_i; i = 1,2,\cdots\}$ be an infinite set of orthonormal elements in an inner product space S with inner product (,). Then the **generalized Fourier series** for some element ϕ is

$$\sum_{i=1}^{\infty} (\phi, g_i)\, g_i \qquad\qquad (3.2.9)$$

Example 3.2.1

Let $\{f_1, f_2,\cdots,f_n\}$ be an orthogonal set of elements in linear space S with inner product (,) and norm $||f||^2 = (f,f)$.

$$\text{Then} \quad \left\| \sum_{i=1}^{n} f_i \right\|^2 = \left(\sum_{i=1}^{n} f_i,\ \sum_{i=1}^{n} f_i \right)$$

$$= \sum_{i=1}^{n} \sum_{i=1}^{n} (f_i, f_j)$$

$$= \sum_{i=1}^{n} (f_i, f_i) = \sum_{i=1}^{n} ||f_i||^2$$

66

Example 3.2.2

The Gram–Schmidt process can be used to normalize the $\mathbf{R}^4$ vectors $\{(1,0,1,0), (1,1,1,1), (-1,2,0,1)\}$ by

$$\hat{g}_1 = \mathbf{v_1} = (1,0,1,0)$$

$$g_1 = \hat{g}_1/||\hat{g}_1|| = (1,0,1,0)/\sqrt{2}$$

$$\hat{g}_2 = \mathbf{v_2} - (\mathbf{v_2}, g_1)\, g_1$$

$$= (1,1,1,1) - (\sqrt{2})\,(1,0,1,0)/\sqrt{2} = (0,1,0,1)$$

$$g_2 = \hat{g}_2/||\hat{g}_2|| = (0,1,0,1)/\sqrt{2}$$

$$\hat{g}_3 = \mathbf{v_3} - (\mathbf{v_3}, g_1)\, g_1 - (\mathbf{v_3}, g_2)\, g_2$$

$$= (-1,1,1,-1)/2$$

$$g_3 = \hat{g}_3/||\hat{g}_3|| = (-1,1,1,-1)/2$$

Example 3.2.3

Let θ be notation for the operator

$$\theta(y) = y'' + y$$

where $y \in C^2\,[0,\pi/2]$, where $C^2\,[0,\pi/2]$ is the linear space of functions twice–differentiable on $\Omega = [0,\pi/2]$. Auxilliary conditions are $y(0) = 0$, $y(\frac{\pi}{2}) = 1$. Set $\theta(y) = 0$ on Ω.

Define the function $(,)$ on $S = C^2\,[0,\pi/2]$ for f and g in S by

$$(f,g) = \int_0^{\pi/2} (\theta f)(\theta g)\, dx + (fg)\big|_0 + (fg)\big|_{\pi/2}$$

Then $(,)$ is an inner product on S. This is demonstrated as follows (for $f,g,$ and $h \in S$):

(i) $(f,f) = \int_0^{\pi/2} (\theta f)^2 dx + f^2\big|_0 + f^2\big|_{\pi/2} \geq 0$

(ii) $(f,f) = 0$ implies $\theta f = 0$ on Ω and $f(0) = f(\pi/2) = 0$. The general solution, y_g, to $\theta y = 0$ is $y = c_1 \sin x + c_2 \cos x$; hence, $(f,f) = 0$ implies f is the zero element in S.

(iii) $(f,g) = (g,f)$ by observation

(iv)   for $c \in \mathbf{R}$, $\theta cf = c\theta f$, and

$$(cf,g) = \int_{0}^{\pi/2} (\theta cf)(\theta g)\, dx + (cfg)\,\big|_0 + (cfg)\big|_{\pi/2}$$

$$= c(f,g)$$

(v) $(f+g,h) = (f,h) + (g,h)$ by observation.

Example 3.2.4

Consider the set S and inner product (,) on S as given in Example 3.2.3. Let S_2 be the subspace of $C^2[0, \pi/2]$ spanned by $\{f_1,f_2\} = \{\sin x, \cos x\}$. Noting that $\theta f_1 = \theta f_2 = 0$, the basis is orthonormalized as follows:

$$||f_1||^2 = (f_1,f_1) = \int_{0}^{\pi/2} \theta f_1 \theta f_1 dx + f_1^2\big|_0 + f_1^2\big|_{\pi/2}$$

$$= \int_{0}^{\pi/2} (0)\, dx + (0) + (1) = 1$$

$$g_1 = f_1/||f_1|| = f_1 = \sin x$$

$$\hat{g}_2 = f_2 - (f_2,g_1)\, g_1 = f_1 - \int_{0}^{\pi/2} \theta f_2 \theta g_1\, dx$$

$$+ (f_2 g_1)\big|_0$$

$$+ (f_2 g_1)\big|_{\pi/2}$$

$$= \cos x - \int_{0}^{\pi/2} (0)\, dx + (0) + (0) = \cos x = f_2$$

$$\hat{g}_2{}^2 = (\hat{g}_2,\hat{g}_2) = (f_2,f_2) = \int_{0}^{\pi/2} \theta f_2 \theta f_2\, dx + f_2^2\big|_0 + f_2^2\big|_{\pi/2}$$

$$= \int_{0}^{\pi/2} (0)\, dx + (1) + (0) = 1$$

$$g_2 = \hat{g}_2/||\hat{g}_2|| = \hat{g}_2 = f_2$$

Thus $\{f_1, f_2\}$ is an orthonormal basis for S_2.

Example 3.2.5

Given that $\{f_1, f_2\} = \{\sin x, \cos x\}$ are an orthonormal basis to S_2 on $[0, \pi/2]$ for the inner product $(,)$ defined in Example 3.2.3 and Example 3.2.4, then the best approximation ($\hat{\phi}_2$ in S_2) of the solution ϕ to the previous $\theta(y) = 0$, $y(0) = 0$, $y(\pi/2) = 1$ is determined using (3.2.7) and (3.2.8) to be

$$\hat{\phi}_2 = (\phi, f_1) \, f_1 + (\phi, f_2) \, f_2$$

where

$$(\phi, f_1) = \int_0^{\pi/2} \theta\phi \, \theta f_1 \, dx + (\phi f_1)\big|_0 + (\phi f_2)\big|_{\pi/2}$$

$$= \int_0^{\pi/2} (0) \, dx + (0) + (1) = 1$$

$$(\phi, f_2) = \int_0^{\pi/2} \theta\phi \, \theta f_2 \, dx + (\phi f_2)_0 + (\phi f_2)_{\pi/2}$$

$$= \int_0^{\pi/2} (0) \, dx + (0) + (0) = 0$$

Thus,

$$\hat{\phi}_2 = f_1 = \sin x$$

which is the problem solution.

Example 3.2.6

A basis for a two-dimensional linear subspace S_2 in $\mathbf{R}^4$ is $\{f_1, f_2\} = \{\mathbf{v_1}, \mathbf{v_2}\}$ where $\mathbf{v_1} = (1, 2, 3, 4)^T$ and $\mathbf{v_2} = (2, 0, 0, 1)^T$. A vector $\mathbf{b} = (1, 2, 1, 2)^T \in \mathbf{R}^4$ is to be approximated by $\hat{\phi}_2 = \lambda_1 \mathbf{v_1} + \lambda_2 \mathbf{v_2}$ using $(,)$ as the dot product. The elements $\{f_1, f_2\}$ are orthonormalized using the defined inner product,

$$\hat{g}_1 = f_1 = \mathbf{v_1}$$

$$g_1 = \hat{g}_1/||\hat{g}_1|| = \mathbf{v_1}/||\mathbf{v_1}||$$

$$||\mathbf{v_1}||^2 = (\mathbf{v_1}, \mathbf{v_1}) = 30; \quad ||\mathbf{v_1}|| = \sqrt{30}$$

$$g_2 = (1,2,3,4)^T/\sqrt{30}$$

$$\hat{g}_2 = f_2 - (f_2, g_1)g_1 = (2,0,0,1)^T - (6/\sqrt{30})(1,2,3,4)^T/\sqrt{30}$$

$$= (9,-2,-3,1)^T/5$$

$$(\hat{g}_2, \hat{g}_2) = 95/25, \quad ||\hat{g}_2|| = \sqrt{95}/5$$

$$g_2 = \hat{g}_2/||\hat{g}_2|| = (9,-2,-3,1)^T/\sqrt{95}$$

The best approximation $\hat{\phi}_2 \in S_2$ of $\mathbf{b}$ is

$$\hat{\phi}_2 = (\mathbf{b}, g_1) g_1 + (\mathbf{b}, g_2) g_2$$

$$= (8/15)(1,2,3,4)^T + (4/95)(9,-2,-3,1)^T$$

The approximation $\hat{\phi}_2$ is rewritten as a linear combination of the original S_2 basis $\{f_1, f_2\} = \{\mathbf{v_1}, \mathbf{v_2}\}$ by resolution of the orthonormalized vectors

$$\hat{\phi}_2 = (8/15) \mathbf{v_1} + (4/95) (5 \mathbf{v_2} - \mathbf{v_1})$$

$$= (28/57) \mathbf{v_1} + (12/57) \mathbf{v_2}$$

which equates to the results of Example 1.8.4. Note that the procedure utilized to find the best approximation is equivalent to analysis of the matrix system

$$\begin{bmatrix} 1 & 2 \\ 2 & 0 \\ 3 & 0 \\ 4 & 1 \end{bmatrix} \begin{Bmatrix} \lambda_1 \\ \lambda_2 \end{Bmatrix} = \begin{Bmatrix} 1 \\ 2 \\ 1 \\ 2 \end{Bmatrix}$$

Similarly for basis $\{f_1, f_2\} = \{w_1, w_2\}$ where w_1 and w_2 are in R^n, and $d \in R^n$, n large, $[w_1 \; w_2] \begin{bmatrix} \lambda 1 \\ \lambda 2 \end{bmatrix} = d$. The best approximation method utilizes such matrix systems for developing the approximation results.

Example 3.2.7

Let constants λ_1, λ_2, and λ_3 be chosen such that $\hat{\phi}_3 = \lambda_1 + \lambda_2 x + \lambda_3 x^2$ is the best approximation to $\phi = \cos x$ on $\Omega = [0, \pi/2]$. Note that successive derivatives of $\hat{\phi}_3$ and ϕ result in poorer approximation accuracy. Consequently, use of approximations to develop higher order derivatives may be inappropriate.

Table 3.1

$\hat{\phi}_3$ Approximation Derivatives vs. ϕ Derivatives

Order of Derivative n	$\dfrac{d^n \hat{\phi}_3}{dx^n}$	$\dfrac{d^n \phi}{dx^n}$
1	$2\lambda_3 x + \lambda_2$	$-\sin x$
2	$2\lambda_3$	$-\cos x$
3	0	$\sin x$
4	0	ϕ
.	.	.
.	.	.
.	.	.

3.3. Approximations in $L_2(E)$

Let $\{\phi_n\}$ be an orthonormal set of functions in E and $\phi \in L_2(E)$. Define an approximation function by the series $\gamma_1 \phi_1 + \gamma_2 \phi_2 + \cdots$. Compute constants γ_j^* by $\gamma_j^* = \int_E \phi \; \phi_j$.

Then the γ_j^* are called the **generalized Fourier coefficients**. Because both ϕ and $\{\phi_n\}$ are in $L_2(E)$, the γ_j^* values exist.

It is noted that as additional functions are added to a finite set of basis functions $\{\phi_n\}$, the generalized Fourier coefficients γ_j^* are still calculated by the integral $\int_E \phi \; \phi_j$, and the mode of convergence being considered is L_p for p = 2.

3.3.1. Parseval's Identity

Let $\{\phi_n\}$ be an orthonormal set of functions in E with generalized Fourier coefficients computed by $\lambda_j = \int_E \phi \; \dot{\phi}_j$. Also let ϕ and $\{\phi_n\}$ be in $L_2(E)$. Then if $\|(\lambda_1 \phi_1 + \lambda_2 \phi_2 + \cdots + \lambda_n \phi_n) - \phi\| \to 0$ as $n \to \infty$, we have L_p convergence with p = 2 and $\int \phi^2 = \sum_{j=1}^{\infty} \lambda_j^2$. This equality is called **Parseval's Identity.**

3.3.2. Bessel's Inequality

It can be shown that $\int_S \phi^2 \geq \sum_{j=1}^{\infty} \lambda_j^2$ which is called Bessel's Inequality. This inequality is used in the Best Approximation Method to evaluate the rate of L_2 convergence.

3.4. Vector Representations and Best Approximations

Consider the set of functions $\{f_j\}$ on the domain E = [0,1].

$$\text{Let } (f_j, f_k) = \int_0^1 f_j(x) \; f_k(x) \; dx$$

$$= \lim_{\substack{\Delta x \to 0 \\ n \to \infty}} \sum_{i=1}^{n} f_j(x_i) \; f_k(x_i) \; \Delta x$$

where $\Delta x = 1/n$, and x_i is a partition point in E; that is, x_i is the ith component of the vector $(0, 1/n, 2/n, \cdots, (n-1)/n)$.

Let $\mathbf{F}_1$ be the vector in $\mathbf{R}^n$ with components $\mathbf{F}_1 = (f_1(x_i); i = 1, 2, \cdots, n)$, and $(\mathbf{F}_j, \mathbf{F}_k)$ be the standard inner-product in $\mathbf{R}^n$ given by $(\mathbf{F}_j, \mathbf{F}_k) = \sum_{i=1}^{n} f_j(x_i) \; f_k(x_i)$. Then from the above,

$$(f_j, f_k) = \lim_{\substack{\Delta x \to 0 \\ n \to \infty}} \Delta x (\mathbf{F}_j, \mathbf{F}_k); \; \mathbf{F}_j \in \mathbf{R}^n, \mathbf{F}_k \in \mathbf{R}^n$$

We will be using $(\mathbf{F}_j, \mathbf{F}_k)$ in the approximation of (f_j, f_k) in those instances where unable to integrate (f_j, f_k) directly. It is useful to note that we are approximating the vector space spanned by the $\{f_j\}$ by use of the vector space, $\mathbf{R}^n$, where n is determined by the level of computational effort invested into the approximation of the integrals used in the inner-product, (f_j, f_k).

We now examine the function $f(x) = e^x$ on $E = [0,1]$.

The approximation problem is to find the best approximation of $f(x) = e^x$ using the basis functions $\{f_1, f_2, f_3\} = \{1, x, x^2\}$.

Vectors $\{F_1, F_2, F_3\}$ can be generated by evaluating each f_i at, for example, five points of evaluation $x = 0, 0.25, 0.50, 0.75, 1.0$, respectively. Then in R^5

$F_1 = (1,1,1,1,1)$

$F_2 = (0, 0.25, 0.50, 0.75, 1.0)$

$F_3 = (0, 1/16, 1/4, 9/16, 1.0)$

and

$F = (1, 1.284, 1.649, 2.117, 2.718)$

where F is $\phi = e^x$ evaluated at the specified points in E. The set of vectors $\{F_j\}$ form a basis of a subspace of R^5, and is orthonormalized by the Gram-Schmidt technique as follows:

$G_1 = F_1 / \|F_1\| = (0.447, 0.447, 0.447, 0.447, 0.447)$

$G_2 = (F_2 - (F_2, G_1) G_1) / \|F_2 - (F_2, G_1) G_1\|$
$\quad = (-0.633, -0.316, 0, 0.316, 0.633)$

and

$G_3 = (F_3 - (F_3, G_1) G_1 - (F_3, G_2) G_2) /$
$\quad \|F_3 - (F_3, G_1) G_1 - (F_3, G_2) G_2\|$
$\quad = (-0.534, -0.265, -0.534, -0.265, 0.534)$

It is verified that $(G_i, G_j) = 1$ for $i = j$ and $(G_i, G_j) = 0$ for $i \neq j$.

Then the generalized Fourier coefficients γ_j^* are given by

$\gamma_1^* = (G_1, F) = 3.919$

$\gamma_2^* = (G_2, F) = 1.351$

$\gamma_3^* = (G_3, F) = 0.203$

Thus the best approximation (in the R^3 subspace spanned by vectors $\{F_1, F_2, F_3\}$) of F is given by $\gamma_1^* G_1 + \gamma_2^* G_2 + \gamma_3^* G_3$.

In order to obtain the γ_j to be associated with the original F_j (and hence, f_j), we resolve the G_i into components.

G_1 can be rewritten as a function of F_1 as

$G_1 = (0.447, 0.447, 0.447, 0.447, 0.447)$
$\quad = 0.447 (1,1,1,1,1)$
$\quad = 0.447 F_1$

G_2 is a function of F_1 and F_2 by

$G_2 = a_1 F_2 + a_2 F_1$

Substituting F_1 and F_2 into this expression, one obtains

$G_2 = a_1 (0, 0.25, 0.5, 0.75, 1.0) + a_2(1,1,1,1,1)$

or

$G_2 = (a_2 \ a_2 + 0.25a_1, \ a_2 + 0.5a_1, \ a_2 + 0.75a_1, \ a_2 + a_1)$

Also,

$G_2 = (-0.633, -0.316, 0, 0.316, 0.633)$

Solving for a_1 and a_2,

$a_2 = -0.633$

$a_1 = 1.266$

Thus,

$G_2 = 1.266 \ F_2 - 0.633 \ F_1$

G_3 is a function of F_1, F_2 and F_3 by

$G_3 = b_1 F_3 + b_2 F_2 + b_3 F_1$

Again substituting F_1, F_2 and F_3 into this expression, we obtain

$G_3 = b_1 (0, 1/16, 1/4, 9/16, 1.0)$

$+ b_2 (0, 0.25, 0.5, 0.75, 1.0)$

$+ b_3 (1,1,1,1,1)$

or

$$G_3 = (b_3, \ b_3 + \frac{b_2}{4} + \frac{b_1}{16}, \ b_3 + \frac{b_2}{2} + \frac{b_1}{4}, \ b_3 + \frac{3b_2}{4} + \frac{9b_1}{16}, \ b_3 + b_2 + b_1)$$

Also

$G_3 = (0.534, -0.265, -0.534, -0.265, 0.534)$

We solve for b_1, b_2 and b_3 from the following equations:

$b_3 = 0.534$

$$b_3 + \frac{b_2}{2} + \frac{b_1}{4} = -0.534$$

$b_3 + b_2 + b_1 = 0.534$

Using Gaussian elimination,

$b_3 = 0.534$

$b_2 = -4.272$

$b_1 = 4.272$

And finally, we resolve G_3 into its F_i components as

$G_3 = 4.272 \ F_3 - 4.272 \ F_2 + 0.534 \ F_1$

The best approximation using the $\mathbf{R}^5$ vector representations of the basis $\{f_j\}$ is

$$\mathbf{F} = \gamma_1 \mathbf{F}_1 + \gamma_2 \mathbf{F}_2 + \gamma_3 \mathbf{F}_3$$
$$= (1)\,\mathbf{F}_1 + (0.841)\,\mathbf{F}_2 + (0.868)\,\mathbf{F}_3$$

Using the λ_j for also the f_j elements, the best approximation $f^*(x)$ in the original space is estimated as

$$f^*(x) = 1 + (0.841)x + (0.868)x^2$$

A comparison between $f^*(x)$ and $f(x)$ is as follows:

Table 3.2 Approximation Results of e^x by the Best Approximation Method

x	$f(x) = e^x$	$f^*(x)$	$f(x) - f^*(x)$
0	1.000	1.000	0.000
0.125	1.1331	1.1187	0.014
0.250	1.2840	1.2645	0.020
0.375	1.4550	1.4374	0.018
0.500	1.6487	1.6375	0.011
0.625	1.8682	1.8647	0.0035
0.750	2.1170	2.119	−0.0020
0.875	2.3989	2.400	−0.0011
1.000	2.7183	2.709	0.0093

From Table 3.2, the tabled maximum relative error occurs at $x = 0.25$ where the relative error is 1.56 percent.

There are two points to consider: (i) the approximation can be improved by increasing the dimension of the vector representation; there is a limit to how well the functions $\{1, x, x^2\}$ can approximate e^x on $E = [0,1]$; and (ii) by increasing the set of basis functions, the approximation can be improved. Both of these two concepts are utilized in the Best Approximation Method. The first error is a numerical quadrature error, whereas the second error is due to the choice of basis functions.

The inner-product integration approximation can be improved by increasing the number of evaluation points; e.g., $x = 0$, 0.1, 0.2, 0.3, 0.4, 0.5, 0.6, 0.7, 0.8, 0.9, 1.0. Then following the above procedures, one obtains the vectors

$$\mathbf{F}_1 = (1,1,1,1,1,1,1,1,1,1,1)$$
$$\mathbf{F}_2 = (0,0.1,0.2,0.3,0.4,0.5,0.6,0.7,0.8,0.9,1)$$
$$\mathbf{F}_3 = (0,0.01,0.04,0.09,0.16,0.25,0.36,0.49,0.64,0.81,1)$$

and

$$\mathbf{F} = (1,1.105,1.221,1.350,1.492,1.649,1.822,2.014,2.226,2.460,2.718)$$

The set $\{\mathbf{F}_j\}$ is orthonormalized by the Gram–Schmidt technique as follows:

$$\mathbf{G}_1 = \mathbf{F}_1/||\mathbf{F}_1|| = (0.302,0.302,0.302,0.302,0.302,0.302,0.302,$$
$$0.302,0.302,0.302,0.302)$$

$$\mathbf{G}_2 = (\mathbf{F}_2 - (\mathbf{F}_2, \mathbf{G}_1)\,\mathbf{G}_1)/||\mathbf{F}_2 - (\mathbf{F}_2, \mathbf{G}_1)\,\mathbf{G}_1||$$
$$= (-0.477,-0.381,-0.286,-0.191,-0.095,0,\ 0.095,0.191,$$
$$0.286,\ 0.381,\ 0.477)$$

and

$$\mathbf{G}_3 = (\mathbf{F}_3 - (\mathbf{F}_3, \mathbf{G}_1)\,\mathbf{G}_1 - (\mathbf{F}_3, \mathbf{G}_2)\,\mathbf{G}_2)/$$
$$||\mathbf{F}_3 - (\mathbf{F}_3, \mathbf{G}_1)\,\mathbf{G}_1 - (\mathbf{F}_3, \mathbf{G}_2)\,\mathbf{G}_2||$$
$$= (0.512,0.205,-0.034,-0.205,-0.307,-0.341,-0.307,-0.205,$$
$$-0.034,0.205,0.512)$$

Then the generalized Fourier coefficients, γ_j^*, are computed (as before) by

$$\gamma_1^* = (\mathbf{G}_1, \mathbf{F}) = 5.746$$

$$\gamma_2^* = (\mathbf{G}_2, \mathbf{F}) = 1.781$$

$$\gamma_3^* = (\mathbf{G}_3, \mathbf{F}) = 0.248$$

We now reverse our analysis technique in order to obtain the γ_j. $\mathbf{G}_1$, $\mathbf{G}_2$, and $\mathbf{G}_3$ can be represented as functions of $\mathbf{F}_1$, $\mathbf{F}_2$ and $\mathbf{F}_3$ as follows:

$$\mathbf{G}_1 = (0.302, 0.302, 0.302, 0.302, 0.302, 0.302, 0.302, 0.302,$$
$$0.302, 0.302, 0.302)$$

or

$$G_1 = 0.302\ \mathbf{F}_1$$

G_2 can be resolved into the vectors $\mathbf{F}_1$ and $\mathbf{F}_2$ by

$$G_2 = a_1\mathbf{F}_2 + a_2\mathbf{F}_1$$
$$= a_1(0, 0.1, 0.2, 0.3, 0.4, 0.5, 0.6, 0.7, 0.8, 0.9, 1.0)$$
$$+ a_2(1, 1, 1, 1, 1, 1, 1, 1, 1, 1, 1)$$

Also,

$$G_2 = (-0.477, -0.381, -0.286, -0.191, -0.095, 0, 0.095, 0.191, 0.286,$$
$$0.381, 0.477)$$

Solving for a_1 and a_2, one obtains

$$a_2 = -0.477$$
$$a_1 = 0.954$$

Thus,

$$G_2 = 0.954\ \mathbf{F}_2 - 0.477\ \mathbf{F}_1$$

G_3 is resolved into vectors $\mathbf{F}_1$, $\mathbf{F}_2$ and $\mathbf{F}_3$ as

$$G_3 = b_1\mathbf{F}_3 + b_2\mathbf{F}_2 + b_3\mathbf{F}_1$$

By substituting $\mathbf{F}_1$, $\mathbf{F}_2$ and $\mathbf{F}_3$ into this expression, one obtains

$$G_3 = b_1(0, 0.01, 0.04, 0.09, 0.16, 0.25, 0.36, 0.49, 0.64, 0.81, 1.0)$$
$$+ b_2 (0, 0.01, 0.2, 0.3, 0.4, 0.5, 0.6, 0.7, 0.8, 0.9, 1)$$
$$+ b_3 (1, 1, 1, 1, 1, 1, 1, 1, 1, 1, 1)$$

Also

$$G_3 = (0.512, 0.205, -0.034, -0.205, -0.307, -0.341, -0.307, -0.205,$$
$$-0.034, 0.205, 0.512)$$

We obtain the following equations:

$$b_3 = 0.512$$
$$0.01\ b_1 + 0.1\ b_2 + b_3 = 0.205$$
$$b_1 + b_2 + b_3 = 0.512$$

Solving for b_1, b_2, and b_3,

$$b_1 = 3.411$$
$$b_2 = -3.411$$
$$b_3 = 0.512$$

Thus,

$$G_3 = 3.411\ \mathbf{F}_3 - 3.411\ \mathbf{F}_2 + 0.512\ \mathbf{F}_1$$

$$\gamma_3{}^* = (g_3, f)$$

$$= \int_0^1 6\sqrt{5}\ (x^2 - x + \tfrac{1}{6})\ e^x\ dx$$

$$= \sqrt{5}\ \int_0^1 (6x^2 - 6x + 1)\ e^x\ dx$$

$$= \sqrt{5}\,[7e - 19]$$
$$= 0.063$$

Thus, the best approximation function, $f^*(x)$, of $f(x)$ is

$$f^*(x) = \gamma_1{}^*\ g_1 + \gamma_2{}^*\ g_2 + \gamma_3{}^*\ g_3$$
$$= 1.718(1) + 0.488(2\sqrt{3}\ x - \sqrt{3}) + 0.063\ (6\sqrt{5}\ (x^2 - x + \tfrac{1}{6}))$$
$$= 1.014 + 0.845\ x + 0.845\ x^2$$

In comparison, using our $\mathbf{R}^5$ vector representation, $f^*(x) = 1 + 0.841\ x + 0.868\ x^2$; and using our $\mathbf{R}^{11}$ vector representation, $f^*(x) = 1.0127 + 0.853\ x + 0.846\ x^2$.

We now apply the above procedure to a more interesting problem. Let the basis functions be $\{1, x, x^2, x^3\}$ with the linear problem defined as

$$\frac{d^2\phi}{dx^2} = -2,\ 0 \leq x \leq 1,\ \phi(x=0) = 1\ \text{and}\ \phi(x=1) = 2$$

So $L = \dfrac{d^2\phi}{dx^2}$, $h = -2$, and ϕ_b is given by the 2 point values at $x=0,1$.

The basis functions are represented by the vectors $\mathbf{F_j} = (F_{ji}(\Gamma), Lf_{ji}(\Omega);\ i = 1, 2, \cdots, n)\ j = 1, 2, 3, 4$, where Γ is the boundary and Ω is the interval interior, and i is the index used to describe the evaluation points.

(i)   5 point case $(\mathbf{R}^5)$

We select our 5 point partition as $x = \{0, 1/4, 1/2, 3/4, 1\}$, and $\mathbf{F_j} = (f_j(x=0),\ f_j(x=1),\ Lf_j(x=\tfrac{1}{4}),\ Lf_j(x=\tfrac{1}{2}),\ Lf_j(x=\tfrac{3}{4}))$. From the selected basis, $Lf_1 = 0$, $Lf_2 = 0$, $Lf_3 = 2$, $Lf_4 = 6x$.

The $\mathbf{F_j}$ are determined to be

$$\mathbf{F_1} = (1, 1, 0, 0, 0)$$
$$\mathbf{F_2} = (0, 1, 0, 0, 0)$$
$$\mathbf{F_3} = (0, 1, 2, 2, 2)$$
$$\mathbf{F_4} = (0, 1, \tfrac{3}{2}, 3, \tfrac{9}{2})$$

Now apply the Gram-Schmidt Procedure to find the orthonormal vectors, G_j.

$$G_1 = F_1/\|F_1\| = (\tfrac{1}{\sqrt{2}}, \tfrac{1}{\sqrt{2}}, 0,0,0)$$

$$\|F_1\|^2 = (F_1, F_1) = 2$$

$$\hat{G}_2 = F_2 - (F_2, G_1)\, G_1 = (0,1,0,0,0) - \tfrac{1}{\sqrt{2}}(\tfrac{1}{\sqrt{2}}, \tfrac{1}{\sqrt{2}}, 0,0,0)$$

$$= (-\tfrac{1}{2}, \tfrac{1}{2}, 0,0,0)$$

$$(F_2, G_1) = \tfrac{1}{\sqrt{2}}$$

$$G_2 = \hat{G}_2/\|\hat{G}_2\| = \sqrt{2}\,(-\tfrac{1}{2}, \tfrac{1}{2}, 0,0,0) = (-\tfrac{1}{\sqrt{2}}, \tfrac{1}{\sqrt{2}}, 0,0,0)$$

$$(\hat{G}_2, \hat{G}_2) = \tfrac{1}{2}$$

$$\hat{G}_3 = F_3 - (F_3, G_1)\, G_1 - (F_3, G_2)\, G_2 = (0,1,2,2,2) - \tfrac{1}{\sqrt{2}}(\tfrac{1}{\sqrt{2}}, \tfrac{1}{\sqrt{2}}, 0,0,0) -$$

$$\tfrac{1}{\sqrt{2}}(\tfrac{1}{\sqrt{2}}, \tfrac{1}{\sqrt{2}}, 0,0,0)$$

$$= (0,0,2,2,2)$$

$$(F_3, G_1) = \tfrac{1}{\sqrt{2}}$$

$$(F_3, G_2) = \tfrac{1}{\sqrt{2}}$$

$$G_3 = \hat{G}_3/\|\hat{G}_3\| = \tfrac{1}{\sqrt{12}}(0,0,2,2,2) = (0,0,\tfrac{1}{\sqrt{3}}, \tfrac{1}{\sqrt{3}}, \tfrac{1}{\sqrt{3}})$$

$$(\hat{G}_3, \hat{G}_3) = 12$$

$$\hat{G}_4 = F_4 - (F_4, G_1)G_1 - (F_4, G_2)G_2 - (F_4, G_3)G_3 = (0,1,\tfrac{3}{2}, 3, \tfrac{9}{2})$$

$$- \tfrac{1}{\sqrt{2}}(\tfrac{1}{\sqrt{2}}, \tfrac{1}{\sqrt{2}}, 0,0,0) - \tfrac{1}{\sqrt{2}}(-\tfrac{1}{\sqrt{2}}, \tfrac{1}{\sqrt{2}}, 0,0,0)$$

$$- 3\sqrt{3}\,(0,0,\tfrac{1}{\sqrt{3}}, \tfrac{1}{\sqrt{3}}, \tfrac{1}{\sqrt{3}}) = (0,0,-\tfrac{3}{2}, 0, \tfrac{3}{2})$$

$$(F_4, G_1) = \tfrac{1}{\sqrt{2}}, \quad (F_4, G_2) = \tfrac{1}{\sqrt{2}}, \quad (F_4, G_3) = 3\sqrt{3}$$

$$G_4 = \hat{G}_4/\|\hat{G}_4\| = \tfrac{\sqrt{2}}{3}(0,0,-\tfrac{3}{2}, 0, \tfrac{3}{2}) = (0,0,-\tfrac{1}{\sqrt{2}}, 0, \tfrac{1}{\sqrt{2}})$$

$$(\hat{G}_4, \hat{G}_4) = \tfrac{9}{2}$$

Note that the solution to our problem, $f(x)$, is also represented in vector form F, by

$$F = (\phi_b\,(x=0),\ \phi_b(x=1),\ h(\tfrac{1}{4}),\ h(\tfrac{1}{2}),\ h(\tfrac{3}{4}))$$

$$= (1,2,-2,-2,-2)$$

So the best approximation of F is given by G_1, G_2, G_3, and G_4 as

$$\sum_{i=1}^{4} \gamma_i{}^* G_i$$

where

$$\gamma_1^* = (\mathbf{F}, \mathbf{G}_1) = \frac{3}{\sqrt{2}}$$

$$\gamma_2^* = (\mathbf{F}, \mathbf{G}_2) = \frac{1}{\sqrt{2}}$$

$$\gamma_3^* = (\mathbf{F}, \mathbf{G}_3) = -\frac{6}{\sqrt{3}}$$

$$\gamma_4^* = (\mathbf{F}, \mathbf{G}_4) = 0$$

Performing back substitution we obtain

$$\gamma_4 = 0$$
$$\gamma_3 = -1$$
$$\gamma_2 = 2$$
$$\gamma_1 = 1$$

And the best approximation to $f(x)$ is

$$f^*(x) = \sum_{i=1}^{4} \gamma_i f_i = 1 + 2x - x^2$$

Note that $Lf^* = -2$, and $f^*(0) = 1$, $f^*(1) = 2$

(ii) We now consider a vector representation in $\mathbf{R}^9$. As before,

$$\mathbf{F}_j = (f_j(x=0),\ f_j(x=1),\ Lf_j(\tfrac{1}{8}),\ Lf_j(\tfrac{1}{4}),\ Lf_j(\tfrac{3}{8}),\ Lf_j(\tfrac{1}{2})\ Lf_j(\tfrac{5}{8}),\ Lf_j(\tfrac{3}{4}),$$
$$Lf_j(\tfrac{7}{8}))$$

where f_i's are the same as previously computed. Thus

$$\mathbf{F}_1 = (1,1,0,0,0,0,0,0,0)$$
$$\mathbf{F}_2 = (0,1,0,0,0,0,0,0,0)$$
$$\mathbf{F}_3 = (0,1,2,2,2,2,2,2,2)$$
$$\mathbf{F}_4 = (0,1,\tfrac{3}{4},\tfrac{3}{2},\tfrac{9}{4},3,\tfrac{15}{4},\tfrac{9}{2},\tfrac{21}{4}\ )$$

From the Gram-Schmidt procedure,

$$\mathbf{G}_1 = \mathbf{F}_1/\|\mathbf{F}_1\| = \frac{1}{\sqrt{2}}\ \mathbf{F}_1 = (\tfrac{1}{\sqrt{2}},\tfrac{1}{\sqrt{2}},\ 0,0,0,0,0,0,0)$$

$$(\mathbf{F}_1, \mathbf{F}_1) = 2$$

$$\hat{\mathbf{G}}_2 = \mathbf{F}_2 - \frac{1}{2}\ \mathbf{G}_1 = (-\tfrac{1}{2},\tfrac{1}{2},0,0,0,0,0,0,0)$$

$$(\mathbf{F}_2, \mathbf{G}_1) = \frac{1}{2}\ (\hat{\mathbf{G}}_2,\hat{\mathbf{G}}_2) = \frac{1}{2}$$

$$\mathbf{G}_2 = \sqrt{2}\ \hat{\mathbf{G}}_2 = (-\tfrac{1}{\sqrt{2}},\tfrac{1}{\sqrt{2}},0,0,0,0,0,0,0)$$

$$\hat{G}_3 = \mathbf{F}_3 - \frac{1}{\sqrt{2}}G_1 - \frac{1}{\sqrt{2}}G_2 = (0,0,2,2,2,2,2,2,2)$$

$$(\mathbf{F}_3, G_1) = \frac{1}{\sqrt{2}}, \ (\mathbf{F}_3, G_2) = \frac{1}{\sqrt{2}}$$

$$(\hat{G}_3, \hat{G}_3) = 28, \ G_3 = \frac{1}{2\sqrt{7}}\hat{G}_3$$

$$G_3 = (0,0,\frac{1}{\sqrt{7}},\frac{1}{\sqrt{7}},\frac{1}{\sqrt{7}},\frac{1}{\sqrt{7}},\frac{1}{\sqrt{7}},\frac{1}{\sqrt{7}},\frac{1}{\sqrt{7}})$$

$$\hat{G}_4 = \mathbf{F}_4 - \frac{1}{\sqrt{2}}G_1 - \frac{1}{\sqrt{2}}G_2 - 3\sqrt{7}\ G_3 = (0,0,-\frac{9}{4},-\frac{6}{4},-\frac{3}{4},0,\frac{3}{4},\frac{6}{4},\frac{9}{4})$$

$$(\mathbf{F}_4, G_1) = \frac{1}{\sqrt{2}}, \ (\mathbf{F}_4, G_2) = \frac{1}{\sqrt{2}}, \ (\mathbf{F}_4, G_3) = 3\sqrt{7}$$

$$(\hat{G}_4, \hat{G}_4) = \frac{63}{4}$$

$$G_4 = \frac{2}{3\sqrt{7}}\hat{G}_4 = (0,0,-\frac{3}{3\sqrt{7}},-\frac{1}{\sqrt{7}},-\frac{1}{2\sqrt{7}},0,\frac{1}{2\sqrt{7}},\frac{1}{7},\frac{3}{2\sqrt{7}})$$

And for the exact solution, $f(x)$,

$$\mathbf{F} = (\phi_b(x=0), \ \phi_b(x=1), \ h(\tfrac{1}{8}), \ h(\tfrac{1}{4}), \ h(\tfrac{3}{8}), \ h(\tfrac{1}{2}), \ h(\tfrac{5}{8}), \ h(\tfrac{3}{4}), \ h(\tfrac{7}{8}))$$

$$= (1,2,-2,-2,-2,-2,-2,-2,-2)$$

Now find the γ_i^*'s:

$$\gamma_1^* = (\mathbf{F}, G_1) = \frac{3}{2}$$

$$\gamma_2^* = (\mathbf{F}, G_2) = \frac{1}{2}$$

$$\gamma_3^* = (\mathbf{F}, G_3) = -\frac{14}{\sqrt{7}}$$

$$\gamma_4^* = (\mathbf{F}, G_4) = 0$$

By back-substitution:

$$\gamma_4 = 0$$
$$\gamma_3 = -1$$
$$\gamma_2 = 2$$
$$\gamma_1 = 1$$

And again, $f^*(x) = 1 + 2x - x^2$, which is the exact solution to our linear operator equation.

3.5. COMPUTER PROGRAM:

We now focus upon systems of linear equations, $\mathbf{Ax} = \mathbf{b}$, where $\mathbf{A}$ is an m×n matrix, $\mathbf{b}$ is a column vector of dimension n, and $\mathbf{x}$ is a column vector of unknown values and is also of dimension n. In further detail, $\mathbf{Ax} = \mathbf{b}$ is expanded as

$$
\begin{bmatrix}
a_{11} & a_{12}\cdots & a_{1n} \\
a_{21} & a_{22}\cdots & a_{2n} \\
\cdot & \cdot & \cdot \\
\cdot & \cdot & \cdot \\
\cdot & \cdot & \cdot \\
a_{m,1} & a_{m,2}\cdots & a_{m,n}
\end{bmatrix}
\begin{pmatrix}
x_1 \\ x_2 \\ \cdot \\ \cdot \\ \cdot \\ x_n
\end{pmatrix}
=
\begin{pmatrix}
b_1 \\ b_2 \\ \cdot \\ \cdot \\ \cdot \\ b_m
\end{pmatrix}
$$

where the a_{ij}, x_i, and b_i are real numbers. The problem is posed where given the a_{ij} and b_i, what are the x_i? Standard texts on linear algebra provide the theorems regarding existence of such solution vectors, x, and it is recalled that there are three solution scenarios: (i) there is no solution vector; (ii) there is a unique solution vector; (iii) there are an infinity of solution vectors.

In our analysis, the matrix equation problem is recast as follows: What is the choice of x_i in x, that minimize the ℓ_2 error in satisfying

$$
x_1
\begin{pmatrix}
a_{11} \\ a_{21} \\ \cdot \\ \cdot \\ \cdot \\ a_{m,1}
\end{pmatrix}
+ x_2
\begin{pmatrix}
a_{12} \\ a_{22} \\ \cdot \\ \cdot \\ \cdot \\ a_{m,2}
\end{pmatrix}
+\cdots+ x_n
\begin{pmatrix}
a_{1n} \\ a_{2n} \\ \cdot \\ \cdot \\ \cdot \\ a_{m,n}
\end{pmatrix}
=
\begin{pmatrix}
b_1 \\ b_2 \\ \cdot \\ \cdot \\ \cdot \\ b_m
\end{pmatrix}
$$

The matrix problem is now solved by the previous procedures with $F = b$, and $F_i = (a_{1i}, a_{2i}, \cdots, a_{m,i})$ for $i=1,2,\cdots,n$. It is seen from the previous discussion that use of vector representations of the selected functions (to be used in the approximation) results in the solution of a matrix system where the matrix A is composed of column vectors of the F_i.

PROGRAM 1. MATRIX SYSTEM SOLVER

A key portion of our approximation strategy is to solve for the best approximation in an equivalent vector space in R^n. As a result, we will need to find the best fit, in a "least square error" sense, in solving for matrix systems. Computer PROGRAM 1 provides source code for such a matrix solver.

Application of PROGRAM 1

Consider the example problem (Section 3.4) of finding the best approximation of $f(x) = e^x$, using a combination of the basis functions from from $\{1, x, x^2\}$. The input data to PROGRAM 1 are as follows:

```
3,5,1
1,1.284,1.649,2.117,2.718
1 1 1 1
0 .25 .5 .75 1.
0 .0625 .25 .5625 1
```

The output PROGRAM 1 area as follows:

```
*** NODAL POINT VECTOR EXPANSION, F(I) ***

.1000D+01   .1000D+01   .1000D+01   .1000D+01   .1000D+01
.0000D+00   .2500D+00   .5000D+00   .7500D+00   .1000D+01
.0000D+00   .6250D-01   .2500D+00   .5625D+00   .1000D+01

*** ORTHOGONAL VECTOR EXPANSION, G(I) ***

 .4472D+00   .4472D+00   .4472D+00   .4472D+00   .4472D+00
-.6325D+00  -.3162D+00  -.1433D-16   .3162D+00   .6325D+00
 .5345D+00  -.2673D+00  -.5345D+00  -.2673D+00   .5345D+00

*** ORTHOGONAL TEST (G(I),G(J)) ***

 I  J   (G(I),G(J))
 1  2     .2683D-17
 1  3     .4762D-16
 2  3    -.3154D-15

*** EVALUATION COEFFICIENTS (G(I),B(I))/(G(I),G(I)) ***

.1754D+01   .1708D+01   .8423D+00

*** BACK SUBSTITUTION COEFFICIENTS ***

.1005D+01   .8653D+00   .8423D+00
```

```
*** BESSEL INEQUALITY ***
 .1724D+02  >=   .1724D+02 AND THE DIFFERENCE IS   .2713D-03
```

```
$STORAGE:2
C
C          MAIN PROGRAM
C
C          THIS IS A GENERALIZED MATRIX SOLVER USING
C          THE MODIFIED GRAM-SCHMIDT ORTHONORMALIZATION PROCESS
C
           IMPLICIT DOUBLE PRECISION(A-H,O-Z)
           COMMON/BLK 1/ F(20,600),Y(600)
           COMMON/BLK 2/ VALUE(600)
           COMMON/BLK 3/ B(20),S(20)
           COMMON/BLK 4/ G(20,600)
           COMMON/BLK 5/ XK(20,20)
C
C          OPEN DATA FILES
C
           NRD=1
           NWT=2
           OPEN(UNIT=NRD,FILE='MGSOP.DAT',STATUS='OLD')
           OPEN(UNIT=NWT,FILE='MGSOP.ANS',STATUS='UNKNOWN')
C
C          INITIALIZE CONSTANTS
C
           SAREA=0.
           EX=0.
           DO 40 I=1,600
           VALUE(I)=0.
           Y(I)=0.
  40       CONTINUE
           DO 50 I=1,600
           DO 50 J=1,20
           F(J,I)=0.
           G(J,I)=0.
  50       CONTINUE
C
C          READ INPUT DATA
C
C.......DEFINITION OF VARIABLES
C
C          NN   : NUMBER OF VECTOR
C          NNOD : NUMBER OF ELEMENT IN EACH VECTOR
C          KODE : 0 - SUMMARY OF RESULTS
C                 1 - DETAIL OF RESULTS
C          F(I,J)     : LEFT-HAND-SIDE VECTORS
C          VALUE(I)   : RIGHT-HAND-SIDE VECTORS
C
           READ(NRD,*)NN,NNOD,KODE
           READ(NRD,*)(VALUE(I),I=1,NNOD)
           READ(NRD,*)((F(I,J),J=1,NNOD),I=1,NN)
C
           DO 300 I=1,NNOD
           SAREA=SAREA+VALUE(I)*VALUE(I)
 300       CONTINUE
C
           NBAS=NN
           DO 35 I=1,NBAS
           DO 35 J=1,NNOD
           G(I,J)=F(I,J)
  35       CONTINUE
           IF(KODE.NE.1)GO TO 20
           WRITE(NWT,24)
  24       FORMAT(/,10X,'*** NODAL POINT VECTOR EXPANSION, F(I) ***'/)
           DO 45 I=1,NBAS
           WRITE(NWT,22)(G(I,J),J=1,NNOD)
  22       FORMAT(10(1X,D10.4))
  45       CONTINUE
```

```
C
C         USE THE MODIFIED GRAM-SCHMIDT ORTHONORMALIZATION PROCESS
C         TO DETERMINE SERIES OF ORTHOGONAL VECTORS
C
 20       DO 320 I=1,NBAS
          SUM=0.
          DO 323 J=1,NNOD
          SUM=SUM+G(I,J)*G(I,J)
323       CONTINUE
          SUM=SQRT(SUM)
          S(I)=SUM
          IF(SUM.LT..0000001)S(I)=0.
          DO 325 J=1,NNOD
          IF(SUM.GT.0.)G(I,J)=G(I,J)/SUM
          IF(SUM.EQ.0.)G(I,J)=0.
325       CONTINUE
          IF(I.EQ.NBAS)GOTO 320
          IP1=I+1
          DO 310 KK=IP1,NBAS
          SUM1=0.
          DO 330 J=1,NNOD
          SUM1=SUM1+G(I,J)*G(KK,J)
330       CONTINUE
          XXK=-1.*SUM1
          IF(S(I).GT.0.)XK(KK,I)=XXK/S(I)
          IF(S(I).EQ.0.)XK(KK,I)=0.
          DO 340 J=1,NNOD
          G(KK,J)=G(KK,J)+XXK*G(I,J)
340       CONTINUE
310       CONTINUE
320       CONTINUE
          IF(KODE.NE.1.)GO TO 55
          WRITE(NWT,25)
25        FORMAT(/,10X,'*** ORTHOGONAL VECTOR EXPANSION, G(I) ***'/)
          DO 47 I=1,NBAS
47        WRITE(NWT,22)(G(I,J),J=1,NNOD)
C.......CHECK ORTHOGONALITY OF VECTORS G(I)
          WRITE(NWT,26)
26        FORMAT(/,10X,'*** ORTHOGONAL TEST (G(I),G(J)) ***',//,
     1    '    I   J    (G(I),G(J))')
55        DO 52 I=1,NBAS
          IP1=I+1
          IF(I.EQ.NBAS)GO TO 80
          DO 70 K=IP1,NBAS
          SUM=0.
          DO 60 J=1,NNOD
          SUM=SUM+G(I,J)*G(K,J)
60        CONTINUE
          IF(KODE.EQ.1)WRITE(NWT,3)I,K,SUM
3         FORMAT(2X,2I3,3X,D10.4)
70        CONTINUE
52        CONTINUE
80        CONTINUE
C.......COMPUTE THE CONFFICIENTS OF B(I)=(VALUE(I),G(I))
          WRITE(NWT,28)
28        FORMAT(/,10X,'*** EVALUATION COEFFICIENTS',
     1    '  (G(I),B(I))/(G(I),G(I)) ***'/)
          SUM=0.
          DO 120 I=1,NBAS
          BK1=0.
          DO 130 J=1,NNOD
          BK1=BK1+VALUE(J)*G(I,J)
130       CONTINUE
C.......COMPUTE THE NORM OF THE GENERALIZED FOURIER COEFFICIENTS
          IF(S(I).GT.0.)B(I)=BK1/S(I)
          IF(S(I).EQ.0.)B(I)=0.
```

```
             SUM=SUM+BK1*BK1
120          CONTINUE
             WRITE(NWT,22)(B(I),I=1,NBAS)
C.......COMPUTE BASIS FUNCTION COEFFICIENTS (BACK-SUBSTITUTION)
660          DO 200 I=NBAS,1,-1
             IF(I.EQ.NBAS)XK(NBAS,I)=B(NBAS)
             IF(I.NE.NBAS)XK(NBAS,I)=XK(NBAS,I)*B(NBAS)+B(I)
200          CONTINUE
             NTOT1=NBAS-1
             DO 210 I=NTOT1,1,-1
             DO 210 J=I,1,-1
             IF(I.EQ.J)GO TO 210
             IF(I.NE.J)XK(NBAS,J)=XK(NBAS,I)*XK(I,J)+XK(NBAS,J)
210          CONTINUE
             WRITE(NWT,29)
29           FORMAT(/,10X,'*** BACK SUBSTITUTION COEFFICIENTS ***'/)
             WRITE(NWT,22)(XK(NBAS,I),I=1,NBAS)
C
C            APPROXIMATION
C
             WRITE(NWT,99)
             DO 470 I=1,NNOD
             VBAR=0.
             DO 410 J=1,NBAS
 410         VBAR=VBAR+F(J,I)*XK(NBAS,J)
             IF(VALUE(I).EQ.0.)XD=-9999.
             IF(VALUE(I).NE.0.)XD=(VALUE(I)-VBAR)/VALUE(I)
             Y(I)=VBAR
C            WRITE(NWT,4)I,VALUE(I),VBAR,XD
4            FORMAT(/,10X,'*** APPROXIMATE VALUES AND ERRORS'
     1       ,' ***',//,6X,'INTERVAL',9X,'EXACT',9X,'APPROXIMATION',5X,
     2       'RELATIVE ERROR',8X,'F(X,Y)',6X,'ESTIMATED F(X,Y)',3X,
     3       'RELATIVE ERROR')
470          CONTINUE
             WRITE(NWT,99)
99           FORMAT(/,120('='))
C.......BESSEL'S INEQUALITY
             DIFF=SAREA-SUM
             IF(ABS(DIFF).LT.0.00001)DIFF=0.
             WRITE(NWT,36)SAREA,SUM,DIFF
36           FORMAT(/,2X,'*** BESSEL INEQUALITY ***',/,2X,D10.4,
     1       ' >= ',D10.4,' AND THE DIFFERENCE IS ',D10.4,/,120('-')/)
C
             STOP
             END
```

The best approximation of $f(x) = e^x$ on $[0,1]$ is

$\hat{\hat{f}}(x) = 0.8423\ x^2 + 0.8653x + 1.005$ which is different from the best approximation function

$\hat{f}(x) = 0.868x^2 + 0.841\ x + 1$ developed in Section 3.4

A comparison between $\hat{\hat{f}}(x)$ and $f(x)$ is as follows:

Table 3.3. Approximation Results of e^x by the Best Approximation Method (Computer estimated function)

x	$f(x) = e^x$	$\hat{\hat{f}}(x)$	$\hat{\hat{f}}(x) - f(x)$
0	1.000	1.005	−0.005
0.125	1.1331	1.1263	0.0068
0.250	1.2840	1.2740	0.0100
0.375	1.4550	1.4479	0.0071
0.500	1.6487	1.6482	0.0005
0.625	1.8682	1.8748	−0.0066
0.750	2.1170	2.1278	−0.0108
0.875	2.3989	2.4070	−0.0081
1.000	2.7183	2.7126	0.0057

From Table 3.3, the tabled maximum errors are much smaller than those shown on Table 3.2.

CHAPTER 4
LINEAR OPERATORS

4.0. Introduction

Many of the problems that occur in computational mechanics are **linear operator** problems. And being linear operator problems, the existence and uniqueness of solutions is oftentimes determinable based upon the type of relationship being studied and the available data that defines the auxilliary conditions. Additionally, many nonlinear problems are approximately solved by assuming the problem to be linear for short periods in time and/or in space.

4.1. Linear Operator Theory

Let V and W be linear spaces each with their own respective definitions for element addition and scalar (field) multiplication. Let T be a mapping or relationship, such as a computer program, with input $v \in V$ and unique output $w \in W$. This relationship is written as

$$T: V \longrightarrow W \qquad (4.1.1.)$$

or

$$T: v \longrightarrow Tv \qquad (4.1.2.)$$

The set $D(T) = \{v \in V: Tv \text{ is defined}\}$ is called the **domain** and the set $R(T) = \{w \in W: w = Tv \text{ for some } v \in V\}$ is called the **range** of T. If $v_1 \neq v_2 \rightarrow Tv_1 \neq Tv_2$, for all v_1 and v_2 in V, then T is **one-to-one**. If $R(T)$ is a subset of W but there is an element $w \in W$ such that $w \notin R(T)$, then T is an **into** mapping; otherwise, it is onto (i.e. $R(T) = W$).

DEFINITION 4.1.1 (Linear Operator)

Let V and W be real linear spaces and T an operator, $T: V \rightarrow W$. Then T is a linear operator if for all v_1 and v_2 in V and scalars $\lambda \in \mathbf{R}$

(i) $T(v_1 + v_2) = T(v_1) + T(v_2) = w_1 + w_2; \ w_1 = T(v_1), \ w_2 = T(v_2)$
(ii) $T(\lambda v_1) = \lambda T(v_1) = \lambda w_1$

It is noted that in DEFINITION 4.1.1, the appropriate V and W linear space operations are being used. For example, λv_1 is not necessarily the same process as λw_1 unless both linear space definitions of scalar multiplication are similar. It is also noted that usually a linear operator is denoted by L.

DEFINITION 4.1.2. (Inverse Linear Operator)

Let $L_1: U \rightarrow V$ and $L_2: V \rightarrow W$ be real linear operators. The **product** $L_2L_1: U \rightarrow W$ is given for all elements $u \in U$ by

$$L_2L_1u = L_2(L_1u) = L_2(L_1(u))$$

Theorem 4.1.1.

Let L_1, L_2, L_3 be linear operators. Then

i) $L_1(L_2L_3) = (L_1L_2)L_3$

ii) $L_1(L_2+L_3) = L_1L_2+L_1L_3$

iii) $(L_1+L_2)L_3 = L_1L_3+L_2L_3$

DEFINITION 4.1.3. (Inverse Operator, L^{-1})

Let L be a linear operator on the real linear spaces L: $V \rightarrow W$ such that L is one-to-one and onto. Then L has an inverse operator L^{-1} whose domain is contained in W.

Under the conditions of DEFINITION 4.1.3, $LL^{-1} = L^{-1}L = I$ where I is the **identity** operator, $Iv = v$ for all $v \in V$.

DEFINITION 4.1.4. (Bounded Linear Operator)

Let L be a linear operator on the real inner product spaces L: $V \rightarrow W$. Then L is **bounded** if there exists a real number k such that for all v_1 and v_2 in V,

$$||Lv_1 - Lv_2||_W \le k||v_1 - v_2||_V \qquad (4.1.3)$$

The notation $||v_1-v_2||_V$ indicates that the inner product in V is used whereas $||Lv_1-Lv_2||_W$ means that the W inner product is used. It is hereafter assumed understood that the appropriate linear space operations and definitions are used in their respective spaces. The notation L(V,W) is used to designate the linear space of all bounded linear operators from a linear space V to linear space W, where element addition and scalar multiplication are defined by

$$(L_1+L_2)v = L_1v+L_2v, \; v \in V$$

$$(4.1.4)$$

$$(\lambda L_1)v = \lambda(L_1v), \; \lambda \in \mathbf{R}$$

Example 4.1.1

A nxn nonsingular matrix $\mathbf{A}$ is a linear transformation from $V = \mathbf{R}^n$ to $W = \mathbf{R}^n$, by T: $V \rightarrow W$, where for $v \in V$, $Tv = \mathbf{A}v$. Because $\mathbf{A}$ is nonsingular, $\mathbf{A}^{-1}$ exists. The domain $D(T)$ is $\mathbf{R}^n$.

Example 4.1.2

Consider the operator T: $\mathbf{R}^2 \rightarrow \mathbf{R}^2$ where for $v \in \mathbf{R}^2$, $Tv = \mathbf{A}v + \mathbf{b}$ where

$$\mathbf{A} = \begin{bmatrix} 1 & 0 \\ 0 & 1 \end{bmatrix} ; \qquad \mathbf{b} = \begin{pmatrix} \pi \\ e \end{pmatrix}$$

Then T is **nonlinear** because for v_1 and v_2 in $\mathbf{R}^2$,

$$Tv_1 + Tv_2 = \mathbf{A}v_1 + \mathbf{A}v_2 + 2\mathbf{b}$$
$$T(v_1 + v_2) = \mathbf{A}(v_1 + v_2) + \mathbf{b} = \mathbf{A}v_1 + \mathbf{A}v_2 + \mathbf{b}$$

and $T(v_1 + v_2) \neq Tv_1 + Tv_2$

Example 4.1.3

Define the **Kernel** of a linear operator L: $V \rightarrow W$ to be the set

$$Ker(L) = \{v \in V: Lv = \theta_w\} \tag{4.1.5}$$

where θ_w is the zero element in real linear space W. Then the zero element $\theta_v \in V$ is an element of Ker(L) because for any $v \in V$

$$L\theta_v = L(0v) = 0Lv = \theta_w .$$

Example 4.1.4

From <u>Example 4.1.3</u>, let v_1 and v_2 be in Ker(L). Then for $\lambda \in \mathbf{R}$,

$$L(v_1 + v_2) = Lv_1 + Lv_2 = \theta_w + \theta_w = \theta_w$$
$$L(\lambda v_1) = \lambda Lv_1 = \lambda \theta_w = \theta_w$$

Thus the Ker(L) is closed under the same linear space operations of element addition and scalar multiplication as its parent linear space, V. Thus, Ker(L) is a linear subspace of V.

92

Example 4.1.5

Find the Ker(L) for L: $\mathbf{R}^3 \to \mathbf{R}^3$ where for $v \in \mathbf{R}$, $Lv = \mathbf{A}v$, where

$$A = \begin{bmatrix} 1 & 0 & 0 \\ 0 & 1 & 0 \\ 0 & 0 & 0 \end{bmatrix}$$

Solution. Let $v = (x_1,x_2,x_3)$. Then $Lv = (x_1,x_2,y)$ where y is any real number. Thus, for $\theta_W = (0,0,0) = \theta_V$,

$$Ker(L) = \{(x_1,x_2,x_3) \in \mathbf{R}^3 : L(x_1,x_2,x_3) = (0,0,0)\}$$

or $\qquad Ker(L) = (0,0,y) = y(0,0,1)$, $y \in \mathbf{R}$.

It is noted that Ker(L) has dim Ker(L) = 1, and a basis for Ker(L) is $\{f_1\} = \{(0,0,1)\}$.

Example 4.1.6

Let V and W be real linear spaces, and L: V→W be a linear operator. Then because $\theta_V \in Ker(L)$, $L\theta_V = \theta_W$.

(i)   If L is one-to-one, then there is no other element $v \in V$ such that $Lv = \theta_W$, and so $Ker(L) = \{\theta_V\}$.

(ii)   Should $Ker(L) = \{\theta_V\}$, then L must be one-to-one because for any two elements v_1 and v_2 in V, $\theta_W = L(\theta_V) = L(v_1-v_2)$ if and only if $v_1 = v_2$.

From parts (i) and (ii), L is one-to-one if and only if $Ker(L) = \{\theta_V\}$. It is noted that $Ker(L) = \{\theta_V\}$ is a requirement for L^{-1} to be defined.

Example 4.1.7

Let V and W be real linear spaces and L be a linear operator L: V→W. Let $\{f_1,f_2,\cdots,f_n\}$ be a basis for V. Then every $v \in V$ can be written as $v = \lambda_1 f_1+\lambda_2 f_2+\cdots+\lambda_n f_n$. Also, $Lv = \lambda_1 Lf_1+\lambda_2 Lf_2+\cdots+\lambda_n Lf_n$.

Example 4.1.8

Let D be operator notation for differentiation. Let $V = C^1[0,\pi]$ and $W = C^0[0,\pi]$. Then D: V→W is a linear operator as shown for v_1 and v_2 in V, $\lambda \in \mathbf{R}$, by

$$D(v_1+v_2) = Dv_1+Dv_2$$
$$D(\lambda v_1) = \lambda Dv_1$$

Thus D performs as a computer program that has an input of single elements from the real linear space $C^1 [0,\pi]$, and generates the output in $C^0 [0,\pi]$.

Example 4.1.9

Let $V = \mathbf{R}^3$ and $W = \mathbf{R}^1$, and L: $V \to W$ by $Lv = L(x_1, x_2, x_3) = x_1 + x_2 + x_3$.

Then

$$L(v_1 + v_2) = L(x_1+y_1, x_2+y_2, x_3+y_3) = (x_1+y_1+x_2+y_2+x_3+y_3)$$
$$= (x_1+x_2+x_3) + (y_1+y_2+y_3) = Lv_1 + Lv_2$$
$$L(\lambda v_1) = L(\lambda x_1, \lambda x_2, \lambda x_3) = (\lambda x_1 + \lambda x_2 + \lambda x_3)$$
$$= \lambda(x_1+x_2+x_3) = \lambda Lv_1$$

Thus L is a linear operator

4.2. Operator Norms

Theorem 4.2.1

Let L:$V \to W$ be a bounded linear operator on the real normed vector spaces V and W. Then the **operator norm** $||L||$ is determined by

$$||L|| = \sup \{ ||Lv||_w / ||v||_v : v \in V, v \neq \theta_v \} \qquad (4.2.1)$$
$$= \sup \{ ||Lv||_w : v \in V, ||v||_v = 1 \}$$

Note that in (4.2.1), $||Lv||_w$ refers to the norm in the liner operator output space, W, whereas $||v||_v$ refers to the norm in the linear operator input space, V.

The bound of a linear operator, L, is of value in the analysis of input error magnification due to the operation of L on an element from the input linear space. The value of the bound or norm of the linear operator depends upon the definition of norm used in the $D(L)$ and $R(L)$ subspaces of respective parent spaces V and W.

Example 4.2.1

Let L: $V \to W$, where $V = \mathbf{R}^2$ and $W = \mathbf{R}^2$, and for $v \in V$, $Lv = \mathbf{A}v$ where

$$A = \begin{bmatrix} 1 & 2 \\ 2 & 1 \end{bmatrix}$$

In V and W the ℓ_1 norm is used where for $v \in V$ and $v = (x_1, x_2)$, and $w \in W$ and $w = (y_1, y_2)$

$$||v||_1 = |x_1| + |x_2|, \quad ||w||_1 = |y_1| + |y_2|$$

Then when $||v||_1 = 1, |x_1| + |x_2| = 1.$

and $\qquad w = Lv = (y_1, y_2) = ((x_1 + 2x_2), (2x_1 + x_2))$

where $\qquad ||Lv|| = |x_1 + 2x_2| + |2x_1 + x_2|$

$$\leq 3(|x_1| + |x_2|)$$

$$= 3 \text{ for } ||v||_1 = 1.$$

Thus $||L|| = \sup \{||Lv||_1 : ||v||_1 = 1\} = 3.$

Example 4.2.2

Using L, V, W, as defined above, let the norms in V and W be ℓ_2. Then

$$||w||_2^2 = (y_1^2 + y_2^2)$$

$$||v||_2^2 = (x_1^2 + x_2^2)$$

Setting $||v||_2^2 = 1$,

$$||Lv||_2^2 = (5(x_1^2 + x_2^2) + 8x_1 x_2)$$

$$= 5 + 8x_1 x_2$$

But when $||v||_2^2 = 1, x_2^2 = 1 - x_1^2$ and

$$||Lv||_2^2 = 5 + 8x_1 \sqrt{1 - x_1^2}$$

Setting $\frac{\partial}{\partial x_1} ||Lv||_2^2 = 0$ gives $x_1^2 = 1/2$, and $x_2^2 = 1/2$.

Thus $||Lv||_2^2$ is maximized when $(x_1^2, x_2^2) = (1/2, 1/2)$,

that is,

$$\max ||Lv||_2^2 = 9, \text{ or } ||L|| = 3$$

Example 4.2.3

Using L, V, W, as defined above, let the norms in V and W be ℓ_∞.

Then $||v||_\infty = 1$ implies that $|x_1| = |x_2| = 1.$

And $||Lv||_\infty = \max \{(|x_1| + 2|x_2|), (2|x_1| + |x_2|)\}$

giving $||v||_\infty = 1$ implies $||Lv||_\infty = \max\{3, 3\} = 3$

Thus $||L|| = 3.$

Example 4.2.4

Let L: V→W where V = $\mathbf{R}^3$, W = $\mathbf{R}^1$.

Let Lv = L(x_1,x_2,x_3) = $x_1 - 2x_2 + 3x_3$ = w $\in$ W. Then for $||v||_\infty = 1$, $||Lv||_\infty$ is maximized when v = (1,-1,1), giving $||L|| = 6$. Had ℓ_∞ been the norm, then $||v||_1 = 1$ implies $|x_1| + |x_2| + |x_3| = 1$, and $||Lv||_1$ is maximized when $|x_3| = 1$. giving $||L|| = 3$.

Example 4.2.5

Let L: V→W where V = C [0,1], W = $\mathbf{R}^1$. Let Lv be defined for v $\in$ V by v = f (x), and

$$w = Lv = \int_0^1 (x^2 + 1) \, f(x) \, dx$$

Then for $||v||_\infty = ||f(x)||_\infty = 1$,

$$||Lv||_\infty = \max \int_0^1 (x^2 + 1) \, f(x) \, dx, \text{ for } -1 \le f(x) \le 1.$$

Since (x^2+1) is positive, $||Lv||_\infty$ is maximized when f(x) = 1, giving

$$||Lv||_\infty = \int_0^1 (x^2 + 1) \, dx = 4/3$$

Thus $||L|| = 4/3$.

Example 4.2.6

Let L_1: V→V, and L_2: V→V. Then a useful theorem is that $||L_1 L_2|| \le ||L_1|| \; ||L_2||$. Thus for bounded linear operators L: V→V, an iteration operator is given by $L^1 v = Lv$, $L^2 v = L \, Lv$, $L^3 v = L(L(Lv))$, and so forth. Then by the stated theorem, $||L^n|| \le ||L||^n$. From Example 4.2.3, $||L|| = 3$. Now consider

$$L^2 = \mathbf{A}^2 = \begin{bmatrix} 5 & 4 \\ 4 & 5 \end{bmatrix}$$

$$\begin{aligned} ||L^2|| &= \sup \{||L^2 v||_\infty : ||v||_\infty = 1\} \\ &= \max\{(5|x_1| + 4|x_2|), 4|x_1| + 5|x_2|)\} \text{ when } |x_1| = |x_2| = 1 \\ &= 9 \end{aligned}$$

Thus $||L^2|| \le ||L||^2 = 3^2$.

Example 4.2.7

Let $L: V \to W$ be a bounded linear operator. Let $<v_n>$ be a sequence of approximations in V such that $v_n \to v$ in the sense that $||v_n - v||_V \to 0$ as $n \to \infty$, then $||Lv_n - Lv||_W = ||L(v_n - v)||_W \le ||L|| \ ||v_n - v||$. The sequence $<v_n>$ is said to converge to v with **order p** if $||v_n - v||_V \le Cn^{-p}$ for C,p, positive reals. Then $<v_n> \to v$ with order p implies that $<Lv_n> \to Lv$ with at least order p in that

$$||Lv_n - Lv||_W \le C ||L|| n^{-p} \qquad (4.2.2)$$

Example 4.2.8

An important linear operator in computational mechanics is numerical integration of a function $f \in C[\alpha,\beta]$ where $L: C[\alpha,\beta] \to \mathbf{R}^1$ and is given by

$$Lf = \int_\alpha^\beta f(x) \, dx \qquad (4.2.3)$$

For the norm $||w||_\infty$ in $\mathbf{R}^1$,

$$||Lf||_\infty = \int_\alpha^\beta (1) \, dx = \beta - \alpha \qquad (4.2.4)$$

Numerical integration approximately finds Lf in (4.2.3) by a sequence of approximations $<\hat{\phi}_n>$ for $f(x)$ in $[\alpha,\beta]$, where

$$||Lf - L\hat{\phi}_n||_\infty \le (\beta-\alpha) ||f - \hat{\phi}_n||_\infty \qquad (4.2.5)$$

Usually $\hat{\phi}_n$ is a n-point quadrature scheme on $[\alpha,\beta]$. Because $||f - \hat{\phi}_n||_\infty$ is equal to the max $|f(x) - \hat{\phi}_n(x)|$ for $x \in [\alpha,\beta]$, then success of L depends on the maximum difference in values between $\hat{\phi}_n(x)$ and $f(x)$.

Examples of suitable operators L are

(i)   Rectangle Rule: $Lf = (\beta-\alpha) f(\alpha)$

(ii)   Midpoint Rule: $Lf = (\beta-\alpha) f((\alpha+\beta)/2)$

(iii)   Trapezoidal Rule: $Lf = (\beta-\alpha)(f(\alpha) + f(\beta))/2$

(iv)   Simpson's Rule: $LF = (\beta-\alpha)(f(\alpha) + 4f((\alpha+\beta)/2) + f(\beta))/6$

4.3. Examples of Linear Operators in Engineering

Several important mathematical relationships used in engineering studies are linear operator equations.

Example 4.3.1 (Laplace Equation)

Steady state heat transfer, groundwater flow, and many other phenomena are described by the Laplace equation. Let L be the well-known Laplace equation in one-dimension, $L(\phi) = \dfrac{d^2\phi}{dx^2}$. To show L is a linear operator:

$$L(\phi_1 + \phi_2) = \frac{d^2}{dx^2}(\phi_1 + \phi_2) = \frac{d^2(\phi_1)}{dx^2} + \frac{d^2(\phi_2)}{dx^2} = L(\phi_1) + L(\phi_2)$$

$$L(\lambda\phi_1) = \frac{d^2}{dx^2}(\lambda\phi_1) = \frac{\lambda d^2(\phi_1)}{dx^2} = \lambda L(\phi_1)$$

Now consider L to be the Laplace equation in two-dimensions,

$$L(\phi) = \frac{\partial^2\phi}{\partial x^2} + \frac{\partial^2\phi}{\partial y^2} :$$

$$L(\phi_1 + \phi_2) = \frac{\partial^2(\phi_1 + \phi_2)}{\partial x^2} + \frac{\partial^2(\phi_1 + \phi_2)}{\partial y^2}$$

$$= \frac{\partial^2\phi_1}{\partial x^2} + \frac{\partial^2\phi_2}{\partial x^2} + \frac{\partial^2\phi_1}{\partial y^2} + \frac{\partial^2\phi_2}{\partial y^2} = L(\phi_1) + L(\phi_2)$$

$$L(\lambda\phi_1) = \frac{\partial^2(\lambda\phi_1)}{\partial x^2} + \frac{\partial^2(\lambda\phi_2)}{\partial y^2} = \frac{\lambda\partial^2\phi_1}{\partial x^2} + \frac{\lambda\partial^2\phi_2}{\partial x^2} = \lambda L(\phi_1).$$

Example 4.3.2 (Diffusion Equation)

Diffusion processes are common in engineering analysis.

Let $L(\phi) = \dfrac{\partial^2 \phi}{\partial x^2} - \dfrac{\partial \phi}{\partial t}$. Then

$$L(\phi_1 + \phi_2) = \frac{\partial^2(\phi_1 + \phi_2)}{\partial x^2} - \frac{\partial(\phi_1 + \phi_2)}{\partial t} = \frac{\partial^2 \phi_1}{\partial x^2} + \frac{\partial^2 \phi_2}{\partial x^2} - \frac{\partial \phi_1}{\partial t} - \frac{\partial \phi_2}{\partial t}$$

$$= L(\phi_1) + L(\phi_2)$$

$$L(\lambda \phi_1) = \frac{\partial^2(\lambda \phi_1)}{\partial x^2} - \frac{\partial(\lambda \phi_1)}{\partial t} = \frac{\lambda \partial^2(\phi_1)}{\partial x^2} - \frac{\lambda \partial(\phi_1)}{\partial t} = \lambda L(\phi_1)$$

Example 4.3.3 (Diffusion Equation in a Nonhomogeneous Domain)

Nonhomogenuity is the general situtation in studying diffusion.

Let $L(\phi) = \dfrac{\partial}{\partial x} K \dfrac{\partial \phi}{\partial x} - C \dfrac{\partial \phi}{\partial t}$ where $K = K(x)$ and $C = C(x)$. Then

$$L(\phi_1 + \phi_2) = \frac{\partial}{\partial x} K \frac{\partial(\phi_1 + \phi_2)}{\partial x} - C \frac{\partial(\phi_1 + \phi_2)}{\partial t}$$

$$= \frac{\partial K}{\partial x} \frac{\partial(\phi_1 + \phi_2)}{\partial x} + K \frac{\partial^2(\phi_1 + \phi_2)}{\partial x^2} - C \frac{\partial \phi_1}{\partial t} - \frac{\partial \phi_2}{\partial t}$$

$$= K \frac{\partial^2 \phi_1}{\partial x} + \frac{\partial K}{\partial x} \frac{\partial \phi_1}{\partial x} - C \frac{\partial \phi_1}{\partial t} + K \frac{\partial^2 \phi_2}{\partial x} + \frac{\partial K}{\partial x} \frac{\partial \phi_2}{\partial x} - C \frac{\partial \phi_2}{\partial t}$$

$$= L(\phi_1) + L(\phi_2)$$

$$L(\lambda\phi_1) = \frac{\partial}{\partial x} K \frac{\partial(\lambda\phi_1)}{\partial x} - C \frac{\partial(\lambda\phi_1)}{\partial t}$$

$$= \frac{\partial K}{\partial x} \frac{\partial(\lambda\phi_1)}{\partial x} + K \frac{\partial^2(\lambda\phi_1)}{\partial x^2} - C\lambda \frac{\partial\phi_1}{\partial t}$$

$$= \lambda K \frac{\partial^2(\phi_1)}{\partial x^2} + \lambda \frac{\partial K}{\partial x} \frac{\partial(\phi_1)}{\partial x} - \lambda C \frac{\partial\phi_1}{\partial t}$$

$$= \lambda L(\phi_1)$$

Example 4.3.4 (An Integral Equation)

Consider the integral equation $g(x) = \int^x \phi(t)dt$.

Then $L(\phi)$ is defined by $L(\phi)(x) = \int_0^x \phi(t)dt$ is a linear operator:

$$L(\phi_1 + \phi_2) = \int_0^x [\phi_1(t) + \phi_2(t)]\,dt = \int_0^x \phi_1(t)dt + \int_0^x \phi_2(t)dt$$

$$= L(\phi_1) + L(\phi_2), \text{ and}$$

$$L(\lambda\phi) = \int_0^x \lambda\phi(t)dt = \lambda \int_0^x \phi(t)dt = \lambda L(\phi).$$

Example 4.3.5 (A Voltera Integral Equation)

In surface runoff hydrology (see Hromadka et al, 1987), the Voltera integral is used in unit hydrograph theory. Let the flow rate $q(t)$ be given by $q(t) = \int_0^t e(t - x)\phi(x)dx$ where $e(t)$ is the effective rainfall (rainfall less losses) at time t, and $\phi(x)$ is the unit hydrograph. Let $L(\ ) = \int_0^t e(t - x)\ (x)dx$.

Then $L(\phi_1 + \phi_2) = \int_0^t e(t - x)[\phi_1(x) + \phi_2(x)]\,dx = L(\phi_1) + L(\phi_2)$.

Also, $L(\lambda\phi) = \int^t e(t - x)\lambda\phi(x)dx = \lambda L(\phi)$.

Example 4.3.6

A typical problem in engineering computer analysis is the solution of the matrix system $\underset{\sim}{A} = L\underset{\sim}{B}$ where $\underset{\sim}{A}$ is an mx1 vector of unknowns, $\underset{\sim}{B}$ is an mx1 vector of constant values, and L is an mxm matrix.

To show L is a linear operator let $\underset{\sim}{\phi}_1$ and $\underset{\sim}{\phi}_2$ be (mx1) vectors; then immediately $L(\underset{\sim}{\phi}_1 + \underset{\sim}{\phi}_2) = L(\underset{\sim}{\phi}_1) + L(\underset{\sim}{\phi}_2)$, and $L(\lambda\underset{\sim}{\phi}_1) = \lambda L(\underset{\sim}{\phi}_1)$. We see that L is a linear operator which operates on elements of S = {(mx1) vectors of real constants}, and for all s $\in$ S, L(s) $\in$ T. But each t $\in$ T is an (mx1) vector of real constants. Then t $\in$ S, and T and S coincide. Thus, L is an operator from a linear space onto itself.

Example 4.3.7

In Example 4.3.6, the output space T coincided with the input space S. Consider the linear operator which maps S$\rightarrow$T where $\phi(x)$ is an element of C [a,b] and $f(x) = \int_a^x \phi(t)dt, \; x \le b$.

Then $L(\phi)$ is defined by $L(\phi)(x) \int_a^x \phi(t)dt$ and L maps C [a,b] $\rightarrow$ C [a,b].

However if $y = \int_a^b \phi(t)dt$, then $L(\phi) = \int_a^b \phi(t)dt$ is a linear operator which maps C [a,b] $\rightarrow$ $\mathbf{R}$.

Example 4.3.8

Let $L(\phi) = \phi\frac{d\phi}{dx}$. Then L is a nonlinear operator as shown by noting

$$L(\phi_1 + \phi_2) = (\phi_1 + \phi_2)\frac{d}{dx}(\phi_1 + \phi_2)$$

$$= \left[\phi_1\frac{d\phi_1}{dx} + \phi_2\frac{d\phi_2}{dx}\right] + \left[\phi_1\frac{d\phi_2}{dx} + \phi_2\frac{d\phi_1}{dx}\right] \neq \phi_1\frac{d\phi_1}{dx} + \phi_2\frac{d\phi_2}{dx}$$

$$= L(\phi_1) + L(\phi_2). \text{ Thus, L is nonlinear as } L(\phi_1 + \phi_2) \neq L(\phi_1) + L(\phi_2).$$

Similarly, $L(\lambda\phi_1) = \lambda\phi_1 \dfrac{d}{dx}(\lambda\phi_1) = \lambda^2\phi_1\dfrac{d\phi_1}{dx} = \lambda^2 L(\phi_1) \neq \lambda L(\phi_1)$.

Thus $L(\lambda\phi_1) \neq \lambda L(\phi_1)$ and L is nonlinear. It is noted that if either of the two linear operator relationships fail, then the operator L is said to be **nonlinear.**

4.4. Superposition

Some approximation problems may be solved as the sum of a solution of a linear problem and a particular solution.

Poisson's equation in two dimensions is

$$\frac{\partial^2\phi}{\partial x^2} + \frac{\partial^2\phi}{\partial y^2} = f(x,y), \ (x,y) \ \epsilon\Omega, \ \Omega \text{ a subspace in } \mathbf{R}^2. \qquad (4.4.1)$$

with $\phi(x,y)$ or derivative conditions (usually normal or tangential), specified on the problem boundary, Γ. Letting $\phi_p(x,y)$ satisfy

$$\frac{\partial^2\phi_p}{\partial x^2} + \frac{\partial^2\phi_p}{\partial y^2} = f(x,y) \text{ in } \Omega \qquad (4.4.2)$$

then a second operator relationship can be approximately solved that is linear,

$$\frac{\partial^2\eta}{\partial x^2} + \frac{\partial^2\eta}{\partial y^2} = 0, \ (x,y) \ \epsilon \ \Omega \qquad (4.4.3)$$

with boundary conditions for η being those for ϕ less the contribution from ϕ_p. Solving for η in $\Omega\cup\Gamma$, the resulting approximation for ϕ is $\hat{\phi}$,

$$\hat{\phi} = \phi_p + \eta \qquad (4.4.4)$$

Example 4.4.1

The differential operator on (α,β) is usually noted by

$$D_x^k f = D^k f = \frac{d^k f}{dx^k}, \ k = 1,2,\cdots$$

where $f \epsilon V = C^k(\alpha,\beta)$. Then D^k is a linear operator by noting for $\lambda \epsilon \mathbf{R}$, f_1 and f_2 in V,

$$D^k(\lambda f) = \lambda D^k f$$
$$D^k(f_1 + f_2) = D^k f_1 + D^k f_2$$

where $k = 0$ is notation for the identify $D^0 f = f$. Composition of differential operators results in $D^i D^j f = D^i(D^j f) = D^{i+j} f$. From the above, polynomial differential operators $P(D)$ are given by

$$P(D) = \sum_{i=0}^{n} a_i D^{n-i} \qquad (4.4.5.)$$

Of key interest is the output of $P(D)$ (e^{rx}). First note that for $r \in \mathbf{R}$,

$$D^k e^{rx} = r^k e^{rx}$$

and
$$P(D) \left[e^{rx} f(x) \right] = e^{rx} P(D+r) f(x)$$

Thus

$$(D-r)^m \left[x^m e^{rx} \right] = m! e^{rx}$$

A **homogeneous** linear differential equation of interest is $P(D) = 0$, whereas a **nonhomogeneous** linear differential equation is $P(D) = f(x)$. If y_p solves $P(D) \left[y_p \right] = f(x)$, then **a complementary** solution y_c can be used to solve for the general solution of $P(D) \, y_c = 0$. Then the solution to $P(D) = f(x)$ is

$$y = y_p + (C_1 y_1 + C_2 y_2 + \cdots + C_n Y_n)$$

where $\{y_1, y_2, \cdots, y_n\}$ are linearly independent functions on (α, β), $\{C_1, \cdots, C_n\}$ are constants of integration, and $P(D)$ is an n-order polynomial differential operator.

For example, consider

$$\frac{d^3 y}{dx^3} - \frac{6d^2 y}{dx^2} + \frac{6dy}{dx} - 6 = e^{\pi x}$$

or in differential operator form

$$(D-3)(D-2)(D-1) = e^{\pi x}$$

Note that operating both sides by $(D-\pi)$,

$$(D-\pi)(D-3)(D-2)(D-1) = (D-\pi) e^{\pi x}$$

and $(D-\pi) e^{\pi x} = D e^{\pi x} - \pi e^{\pi x} = 0$.

Letting $y_p = \gamma e^{\pi x}$, and substituting y_p into the original expression, $\gamma = (\pi^3 - 6\pi^2 + 6\pi - 6)^{-1}$. Now having y_p, solve the linear operator equation

$$Ly = P(D)y = (D-3)(D-2)(D-1)y = 0$$

where e^{3x}, e^{2x}, e^x satisfy $(D-3)y = 0$, $(D-2)y = 0$, and $(D-1)y = 0$, respectively. The solution then is

$$y = \gamma e^{\pi x} + C_1 e^x + C_2 e^{2x} + C_3 e^{3x}, \quad (\gamma \text{ per above}).$$

Example 4.4.2

Consider the relationship on Ω, a subspace of $\mathbf{R}^2$,

$$\frac{\partial^2 \phi}{\partial x^2} + \frac{\partial^2 \phi}{\partial y^2} = k$$

with sufficient auxilliary conditions, ϕ_b, defined on the boundary Γ. Then a particular solution, y_p, is

$$y_p = \frac{k}{4}(x^2 + y^2)$$

Let $\phi_b^* = \phi_b - y_p$ on Γ. Then a linear operator relationship to be solved is

$$L\phi^* = \frac{\partial^2 \phi^*}{\partial x^2} + \frac{\partial^2 \phi^*}{\partial y^2} = 0$$

with auxilliary conditions $\phi_b^* = \phi_b - y_p$. Solving for ϕ^*, an approximation for ϕ is $\phi = y_p + \phi^*$.

Example 4.4.3

Let $\{f_i\}$ be a set of linearly independent functions on domain Ω with boundary Γ. Let L be a bounded linear operator $L: V \to W$. Let S_n be the subspace of V spanned by the basis $\{f_i\}$. An element $s \in S_n$ is written as

$$s = \lambda_1 f_1 + \lambda_2 f_2 + \cdots + \lambda_n f_n$$

Then

$$Ls = \lambda_1 L f_1 + \lambda_2 L f_2 + \cdots + \lambda_n L f_n$$

where Ls is in W. $R(L)$ is the subspace of W spanned by $\{Lf_1, Lf_2, \cdots, Lf_n\}$.

CHAPTER 5
THE BEST APPROXIMATION METHOD

5.0. Introduction

Many important engineering problems fall into the category of being linear operators, with supporting boundary conditions. In this chapter, an inner-product and norm is used which enables the engineer to approximate such engineering problems by developing a generalized Fourier series. The resulting approximation is the "best" approximation in that a least-squares (L_2) error is minimized simultaneously for fitting both the problem's boundary conditions and satisfying the linear operator relationship (the governing equations) over the problem's domain (both space and time). Because the numerical technique involves a well-defined inner product, error evaluation is readily available using Bessel's inequality. Minimization of the approximation error is subsequently achieved with respect to a weighting of the inner product components, and the addition of basis functions used in the approximation.

5.1. An Inner Product for the Solution of Linear Operator Equations

The general setting for solving a linear operator equation with boundary values by means of an inner product is as follows: Let Ω be a region in R^m with boundary Γ and denote the closure of Ω by $cl(\Omega)$. Consider the real Hilbert space $L_2(cl(\Omega),\ d\mu)$, which has inner product $(f,g) = \int fg\,d\mu$. To construct the inner product for the development of a generalized Fourier Series is to choose the measure μ correctly; that is let μ be one measure μ_1 on Ω and another measure μ_2 on Γ. One choice for a plane region would be for μ_1 to be the usual two dimensional Lebesque measure $d\Omega$ on Ω and for μ_2 to be the usual arc length measure $d\Gamma$ on Γ. Then an inner product is given by

$$(f,g) = \int_{\Omega} fg \ d\Omega + \int_{\Gamma} fg \ d\Gamma \qquad (5.1.1)$$

Consider a boundary value problem consisting of an operator L defined on domain D(L) contained in $L_2(\Omega)$ and mapping into $L_2(\Omega)$, and a boundary condition operator B defined on a domain D(B) in $L_2(\Omega)$ and mapping it into $L_2(\Gamma)$. The domains of L and B have to be chosen so at least for f in D(L), Lf is in $L_2(\Omega)$, and for f in D(B), Bf is in $L_2(\Gamma)$. For example we could have $Lf = \nabla^2 f$, and Bf(s) equal the almost everywhere (ae) radial limit of f at the points on Γ, with appropriate domains.

The next step is to construct an operator T mapping its domain $D(T) = D(L) \cap D(B)$ into $L_2(cl(\Omega))$ by

$$Tf(x) = Lf(x) \text{ for x in } \Omega \left.\begin{matrix} \\ \\ \\ \end{matrix}\right\}$$

$$Tf(s) = Bf(s) \text{ for s on } \Gamma. \tag{5.1.2}$$

From (5.1.2), there exists a single operator T on the Hilbert space $L_2(cl(\Omega))$ which incorporates both the operator L and the boundary conditions B, and which is linear if both L and B are linear.

Consider the inhomogeneous equation $Lf = g_1$ with the inhomogeneous boundary conditions $Bf = g_2$. Then define a function g on $cl(\Omega)$ by $g = g_1$ on Ω and $g = g_2$ on Γ. Then if the solution exists for the operator equation $Tf = g$, the solution f satisfies $\nabla^2 f = g_1$ on Ω, $f = g_2$ on Γ in the usual sense of meaning that the radial limit of f is g_2 on Γ. One way to attempt to solve the equation $Tf = g$ is to look at a subspace D_n of dimension n, which is contained in D(T), and to try to minimize $||Th - g||$ over all the h in D_n.

In this chapter, the mathematical development of the Best Approximation Method is presented. Detailed example problems are included to illustrate the inner products employed in the method, and to demonstrate the progression of steps used in the development of the associated computer program. Extension of the Best Approximation Method to a computer program for the approximation of boundary value problems of the two–dimensional Laplace equation is contained in Chapter 6. Generalization of the computer program to other linear operator problems is the focus of other sections.

5.2. Definition of Inner Product and Norm

Given a linear operator relationship

$$L(\phi) = h \text{ on } \Omega, \quad \phi = \phi_b \text{ on } \Gamma \tag{5.2.1}$$

defined on the problem domain with auxilliary conditions of $\phi = \phi_b$ on the boundary Γ (see Fig. 5.1). Here Ω may represent both time and space, and ϕ_b may be both initial and boundary conditions. It is assumed that the working space is sufficiently restricted (see following) such that ϕ is a unique almost everywhere (ae) solution to (5.2.1).

Choose a set of m linearly independent functions $<f_j>^m$, and let S_m be the m-dimensional space spanned by the elements of $<f_j>^m$. Here, the elements of $<f_j>^m$ will be assumed to be functions of the dependent variables appearing in (5.2.1)

An inner-product is defined for elements of S_m by (u,v) where for $u, v \in S_m$

$$(u,v) = \int_\Gamma uv d\Gamma + \int_\Omega LuLv d\Omega \tag{5.2.2}$$

It is seen that (u,v) is indeed an inner-product, because for elements u, v, w in S_m

(i) $(u,v) = (v,u)$

(ii) $(ku,v) = k(u,v)$, for L a linear operator

(iii) $(u+v,w) = (u,w) + (v,w)$ for L a linear operator

(iv) $(u,u) = \int_\Gamma (u)^2 \, d\Gamma + \int_\Omega (Lu)^2 \, d\Omega \geq 0$

(v) $(u,u) = 0 \Rightarrow u = 0$ ae on Γ, and $Lu = 0$ ae over Ω

The above restrictions on the operator L implies that L is linear (see (ii) and (iii) in the above definition); if $Lu = 0$ ae over Ω and $u = 0$ ae on Γ, this must imply that the solution $u = [0]$, where $[0]$ is the zero element over $\Omega \cup \Gamma$; and for the inner-product to exist, the integrals must be finite. Additionally, each element $u \in S_m$ must satisfy $\int_\Gamma u^2 d\Gamma < \infty$.

For the above restrictions of L and the space S_m, the inner-product is defined and a norm "$|| \ ||$" immediately follows,

$$||u|| \equiv (u,u)^{\frac{1}{2}} \tag{5.2.3}$$

The generalized Fourier series approach can now be used to obtain the "best" approximation $\phi_m \in S_m$ of the function ϕ using the defined inner-product and corresponding norm.

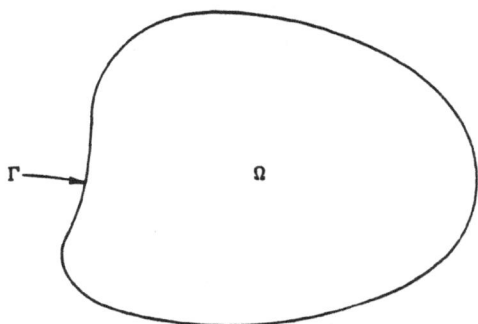

Fig. 5.1. Definition of Problem Domain, Ω,
and Boundary, Γ. (Note: ϕ_b
can include the temporal term
boundary of the intial condition
specification.)

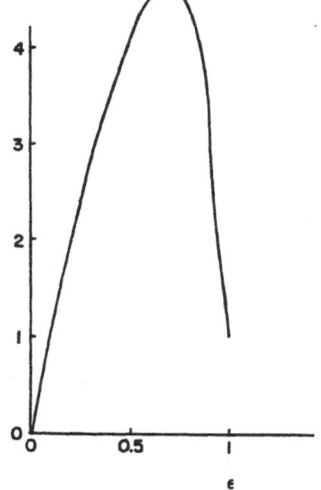

χ_ϵ

Fig. 5.2. Best Approximation Unit
Hydrograph (dashed) and Exact
Unit Hydrograph (solid line).

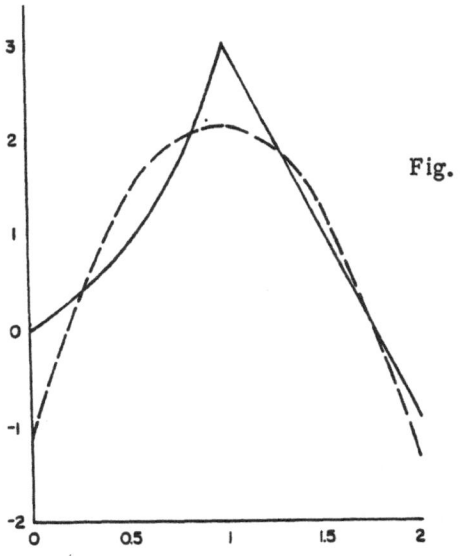

Fig. 5.3. $\chi_\epsilon = (\phi,\phi) - \gamma_1^{*2}$ for
Example Problem 5.5.1.

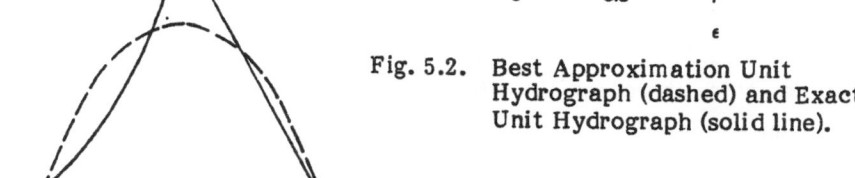

The next step in developing the generalized Fourier series is to construct a new set of functions $<g_j>^m$ which are the orthonormal representation of the $<f_j>^m$.

5.3. Generalized Fourier Series

The functions $<g_j>^m$ can be obtained by the well-known Gram–Schmidt procedure using the defined norm of (5.2.2). That is,

$$g_1 = f_1/||f_1||$$

$$\vdots \qquad\qquad\qquad\qquad\qquad (5.3.1)$$

$$g_m = [f_m - (f_m, g_1)g_1 - \cdots - (f_m, g_{m-1})g_{m-1}]/$$
$$||f_m - (f_m, g_1)g_1 - \cdots - (f_m, g_{m-1})g_{m-1}||$$

Hence, the elements of $<g_j>^m$ satisfy the convenient properties that

$$(g_j, g_k) = \begin{cases} 0, & \text{if } j \neq k \\ \\ 1, & \text{if } j = k \end{cases} \qquad (5.3.2)$$

In the subsequent section, a simple one-dimensional problem illustrates the orthonormalization procedure of (5.3.1).

The elements $<g_j>^m$ also form a basis for S_m but, because of (5.3.2), can be directly used in the development of a generalized Fourier series where the computed coefficients do not change as the dimension m of $<g_j>^m$ increases. That is, as the number of orthonormalized elements increases in the approximation effort, the previously computed coefficients do not change. Each element $\phi_m \in S_m$ can now be written as

$$\phi_m = \sum_{j=1}^{m} c_j g_j, \quad \phi_m \in S_m \qquad (5.3.3)$$

where c_j are unique real constants, that will ultimately be computed.

The ultimate objective is to find the element $\phi_m \in S_m$ such that $||\phi_m - \phi||$ is a minimum. That is, we want $||\phi_m - \phi||^2$ to be a minimum, where

$$||\phi_m - \phi||^2 = \int_{\Gamma} \left(\sum_{j=1}^{m} c_j g_j - \phi_b \right)^2 d\Gamma + \int_{\Omega} \left(L \sum_{j=1}^{m} c_j g_j - L\phi \right)^2 d\Omega \qquad (5.3.4)$$

Recalling that L is a linear operator, and $L\phi = h$ by the problem definition of (5.2.1), then (5.3.4) can be rewritten as

$$\|\phi_m - \phi\|^2 = \int_\Gamma \left(\sum_{j=1}^m c_j g_j - \phi_b \right)^2 d\Gamma + \int_\Omega \left(\sum_{j=1}^m c_j L g_j - h \right)^2 d\Omega \qquad (5.3.5)$$

Thus, minimizing $\|\phi_m - \phi\|^2$ is equivalent to minimizing the error of approximating the boundary conditions and the error of approximating the governing operator relationship in a least-square (or L_2) sense. Because the $<g_j>^m$ are orthonormalized and the inner-product $(\,,\,)$ is well-defined, the coefficients c_j of (5.3.3) are determined to be the generalized Fourier constants, γ_j^*, where

$$\gamma_j^* = (g_j, \phi), \ j = 1, 2, \cdots, m \qquad (5.3.6)$$

Thus

$$\hat{\phi}_m = \sum_{j=1}^m \gamma_j^* g_j = \sum_{j=1}^m (g_j, \phi) g_j \qquad (5.3.7)$$

is the "best" approximation of ϕ, in the space S_m.

Example 5.3.1

To illustrate the previous development, the one-dimensional torsion problem is once more studied. In this example, four polynomials (linearly independent functions) are used as a basis, and we will develop generalized Fourier coefficients by direct integration, rather than by use of a $\mathbf{R}^n$ vector representation. The basis functions considered are

$$<f_j>^4 = <1,\ x,\ x^2,\ x^3>$$

and the differential equation is given by

$$\frac{d^2\phi}{dx^2} = -2, \ \phi(x = 0) = 1 \text{ and } \phi(x = 1) = 2 \text{ for } 0 \le x \le 1.$$

Here $L = \dfrac{d^2}{dx^2}$, $h = -2$, and ϕ_b is given by the two point values at $x = 0$ and 1. The inner-product of (5.2.2) is given as

$$(u,v) = \int_{\Gamma} uvd\Gamma + \int_{\Omega} LuLvd\Omega = uv\Big|_{x=0} + uv\Big|_{x=1} + \int_{\Omega} \frac{d^2(u)}{dx^2}\frac{d^2(v)}{dx^2}d\Omega$$

The 4-dimensional space S_4 is the set of all functions (polynomial) such that $\phi_4(x) = c_1 + c_2 x + c_3 x^2 + c_4 x^3$ where the c_i are real constants.

The orthonormalization of the $<f_j>^4$ proceeds as follows:

For element g_1:

$$(f_1,f_1) = (1)(1)\Big|_{x=0} + (1)(1)\Big|_{x=1} + \int_{x=0}^{1} \frac{d^2(1)}{dx^2}\frac{d^2(1)}{dx^2}dx = 2$$

and

$$g_1 = f_1/||f_1|| = 1/\sqrt{2} = \sqrt{2}/2$$

For element g_2:

$$(f_2,g_1) = (x,\sqrt{2}/2) = (x)(\sqrt{2}/2)\Big|_{x=0} - (x)(\sqrt{2}/2)\Big|_{x=1} + \int_0^1 \frac{d^2(x)}{dx^2}\frac{d^2(\sqrt{2}/2)}{dx^2}dx$$

$$= \sqrt{2}/2$$

We introduce an intermediate function $\hat{g}_2$ by

$$\hat{g}_2 = f_2 - (f_2,g_1)g_1 = x - (\sqrt{2}/2)(\sqrt{2}/2) = x - 1/2$$

$$(\hat{g}_2,\hat{g}_2) = (x-1/2)(x-1/2)\Big|_{x=0} + (x-1/2)(x-1/2)\Big|_{x=1} + \int_0^1 \frac{d^2(x-1/2)}{dx^2}\frac{d^2(x-1/2)}{dx^2}dx$$

$$= 1/2$$

$$\therefore g_2 = \hat{g}_2/||\hat{g}_2|| = (x-1/2)/(\sqrt{2}/2) = (2x-1)/\sqrt{2}$$

Similarly for element g_3:

$$g_3 = (x^2 - x)/2$$

Element g_4:

$$(f_4, g_1) = \sqrt{2}, 2 \; ; \quad (f_4, g_2) = \sqrt{2}/2 \; ; \quad (f_4, g_3) = 3$$

$$\hat{g}_4 = x^3 - (f_4, g_1)g_1 - (f_4, g_3)g_3 = x^3 - \frac{3}{2}x^2 + \frac{x}{2}$$

$$(\hat{g}_4, \hat{g}_4) = \left[x^3 - \frac{3}{2}x^2 + \frac{x}{2}\right]^2 \bigg|_{x=0} + \left[x^3 - \frac{3}{2}x^2 + \frac{x}{2}\right]^2 \bigg|_{x=1} + \int_0^1 \left[\frac{d^2(x^3 - \frac{3}{2}x^2 + \frac{x}{2})}{dx^2}\right]^2 dx$$

$$= 0 + 0 + \int_0^1 (6x - 3)^2 \, dx = 3$$

$$\therefore g_4 = \hat{g}_4 / ||\hat{g}_4|| = (x^3 - \frac{3}{2}x^2 + \frac{1}{2}x)/\sqrt{3}$$

Hence, the orthonormal vectors $\langle \hat{g}_j \rangle^4$ are

$$\langle g_j \rangle^4 = \langle \sqrt{2}/2, \; (2x-1)/\sqrt{2}, \; (x^2 - x)/2, \; (2x^3 - 3x^2 + x)/2\sqrt{3} \rangle$$

Now, any element $\phi_4 \in S_4$ is of the form

$$\phi_4 = \sum_{j=1}^{4} c_j g_j$$

The norm $||\phi_4 - \phi||$ is a minimum when $c_j = \gamma_j^*$ where γ_j^* are the generalized Fourier series coefficients determined from
$$\gamma_j^* = (g_j, \phi).$$
That is,

$$\gamma_j^* = \int_\Gamma g_j \phi_b d\Gamma + \int_\Omega Lf_j \, L\phi \, d\Gamma = \int_\Gamma g_j \phi_b \, d\Gamma + \int_\Omega Lg_j h \, d\Omega$$

where the Lg_j are given by

$$\langle Lg_j \rangle^4 = \langle 0, \; 0, \; 1, \; (6x-3)/\sqrt{3} \rangle$$

Remembering that $h = -2$ by the problem definition, we solve for the γ_j^* as follows:

$$\gamma_1^* = (g_1, \phi) = \left(\frac{\sqrt{2}}{2}\right) \left[\phi_b\right]\Big|_{x=0} + \left(\frac{\sqrt{2}}{2}\right) \left[\phi_b\right]\Big|_{x=1} + 0 = 3\sqrt{2}/2$$

$$\gamma_2^* = (g_2, \phi) = \left(\frac{2x-1}{\sqrt{2}}\right) \left[\phi_b\right]\Big|_{x=0} + \left(\frac{2x-1}{\sqrt{2}}\right) \left[\phi_b\right]\Big|_{x=1} + 0 = \sqrt{2}/2$$

$$\gamma_3^* = (g_3, \phi) = 0 + 0 + \int_0^1 Lg_3 f dx = \int_0^1 (1)(-2)\ dx = -2$$

$$\gamma_4^* = (g_4, \phi) = 0 + 0 + \int_0^1 Lg_4 f dx = \int_0^1 \left(\frac{6x-3}{\sqrt{3}}\right) (-2)\ dx = 0$$

Thus, the best approximation in S_4 is given by

$$\hat{\phi}_4 = \sum_{j=1}^{4} \gamma_j^* g_j = 1 + 2x - x^2$$

It is readily seen that $L\hat{\phi}_4 = -2 = h$, and $\hat{\phi}_4$ satisfies the problem boundary conditions.

Example 5.3.2

A Voltera integral equation (such as occurs in developing transfer functions from watershed rainfall-runoff data) is considered where $q(t)$ is catchment runoff, $\phi(t)$ is a transfer function, t is storm time, and $i(t)$ is effective rainfall,

$$q(t) = \int_0^t i(t-5)\ \phi(s)\ ds,\ 0 \le t \le 2$$

In this example, the effective rainfall intensity (rainfall less that portion retained on the soil, infiltration, etc.) is given by the constant value

$$i(t) = 1,\ 0 \le t \le 2$$

and the runoff hydrograph flowrate q(t) is given by (in units of volume/time)

$$q(t) = \begin{cases} t^3, & 0 \le t \le 1 \\ \\ -2t^2 + 7t - 4, & 1 \le t \le 2 \end{cases}$$

The operator, $L: V \to W$ (for appropriate spaces, V, W) where for $v \in V$,

$$Lv = \int_0^t i(t-s)\, v(s)\, ds$$

L is seen to be a linear operator by noting that for v_1 and v_2 in V and $\lambda \in \mathbf{R}$

$$L(v_1 + v_2) = \int_0^t i(t-s)(v_1(s) + v_2(s))\, ds = \int_0^t (i(t-s)\, v_1(s) + i(t+2)\, v_2(s))\, ds$$

$$= \int_0^t i(t-s)\, v_1(s)\, ds + \int_0^t i(t-s)\, v_2(s)\, ds$$

$$= Lv_1 + Lv_2$$

$$L(\lambda v_1) = \int_0^t i(t-s)\, \lambda v_1(s)\, ds = \lambda \int_0^t i(t-s)\, v_1(s)\, ds = \lambda Lv_1$$

Certain conditions are required between $q(t) \ge 0$ and $i(t) \ge 0$, such as $i(t)$ becomes nonzero at or prior in time that $q(t)$ becomes nonzero, among others.

In this class of problem, neither boundary nor initial conditions are involved, hence the inner product (5.2.2) is

$$(u,v) = \int_\Omega LuLv\, d\Omega$$

(5.3.8)

$$= \int_{t=0}^2 \left[\int_0^t i(t-s)u(s)\, ds \int_0^t i(t-s)v(s)\, ds \right] dt$$

By assumption, i(t−s) = 1 for $0 \leq t \leq 2$, and the inner product reduces to

$$(u,v) = \int_{t=0}^{2} \left[\int_{0}^{t} u(s)\, ds \int_{0}^{t} v(s) ds \right] dt$$

Three elements are considered for basis functions $<f_j>^3$, namely the polynomials $<1, s, s^2>$. The orthonormalized elements $<g_j>^3$ are determined in the following:

g_1:

$$Lf_1 = \int_{0}^{t} (1) ds = t$$

$$\therefore (f_1, f_1) = \int_{0}^{2} t^2 dt = 8/3; \quad ||f_1|| = 2\sqrt{\frac{2}{3}}$$

and $g_1 = f_1 / ||f_1|| = \sqrt{\frac{3}{8}}$

g_2:

$$Lf_2 = \int_{0}^{t} s\, ds = t^2/2$$

$$Lg_1 = \int_{0}^{t} \sqrt{\frac{3}{8}}\, ds = t\sqrt{\frac{3}{8}}$$

$$\therefore (f_2, g_1) = \int_{0}^{2} Lf_2 Lg_1 dt = \int_{0}^{2} \left(\frac{t^2}{2} \right) \left(t\sqrt{\frac{3}{8}} \right) dt = \sqrt{\frac{3}{2}}$$

Now $\hat{g}_2 = f_2 - (f_2, g_1)g_1 = s - 3/4$

$$L\hat{g}_2 = \int_0^t (s - 3/4)ds = \frac{t^2}{2} - \frac{3t}{4}$$

$$\therefore (\hat{g}_2, \hat{g}_2) = \int_0^2 \left(\frac{t^2}{2} - \frac{3t}{4} \right)^2 dt = \frac{1}{10}$$

$$\therefore g_2 = \hat{g}_2 / \| \hat{g}_2 \| = \left(s - \frac{3}{4} \right) \sqrt{10}$$

g_3:

Analogous to the above,

$$(f_3, g_1) = \frac{16}{5} \sqrt{\frac{1}{6}}$$

$$(f_3, g_2) = \sqrt{10}/5.625$$

$$\therefore \hat{g}_3 = f_3 - (f_3, g_1)g_1 - (f_3, g_2)g_2$$

$$= s^2 + 0.5\bar{3} - 1.\bar{7}s$$

where the overbar notation indicates repetitive digits. Finally,

$$g_3 = \hat{g}_3 / \| \hat{g}_3 \| = 10.5234s^2 - 18.708s + 5.6125$$

The generalized Fourier coefficients are determined as before by

$$\gamma_1^* = (g_1, \phi) = \int_\Omega Lg_1 L\phi d\Omega$$

$$= \sqrt{\frac{3}{8}} \int_0^1 (t)(t^3)dt + \sqrt{\frac{3}{8}} \int_1^2 (t)(-2t^2 + 7t-4)dt = 1.85$$

$$\gamma_2^* = (g_2, \phi) = 0.21082$$

$$\gamma_3^* = (g_3, \phi) = -0.325$$

Thus the best approximation is developed (for the defined inner product of (5.3.8)) by

$$\hat{\phi}_3 = 3.42s^2 + 6.7467s - 1.1865$$

For this example problem, the exact solution is determined by taking the derivative of the q(t) function (rewritten in terms of the variable s)

$$\phi(s) = \begin{cases} 3s^2, & 0 \leq s \leq 1 \\ \\ -4s+7, & 1 \leq s \leq 2 \end{cases}$$

Figure 5.2 compares the exact solution $\phi(s)$ to the approximation function $\hat{\phi}_3(s)$ developed from using only 3 polynomial basis functions.

It is noted that although the Example 5.3.1 and Example 5.3.2 are different operator relationships (i.e., a differential equation and a Voltera integral), the approximation method and procedures are identical.

5.4. Approximation Error Evaluation

Due to the generalized Fourier series approach and the definition of the inner-product, Bessel's inequality applies. That is, for any dimension m

$$(\phi,\phi) \geq \sum_{j=1}^{m} (gj,\phi)2 = \sum_{j=1}^{m} \gamma_j^{*2} \tag{5.4.1}$$

where

$$(\phi,\phi) = \int_{\Gamma} (\phi)^2 d\Gamma + \int_{\Omega} (L\phi)^2 d\Omega = \int_{\Gamma} \phi^2 d\Gamma + \int_{\Omega} h^2 d\Omega \tag{5.4.2}$$

Equation (5.4.2) is readily evaluated and forms an upper bound to the sum of $(gj,\phi)^2$ as the dimension m increases. Consequently, one may interact with the approximation effort by carefully adding basis functions to the $<fj>^m$ in an effort to best reduce the difference computed by Bessel's inequality.

For __Example 5.3.1__, the problem definition provides

$$(\phi,\phi) = \phi^2\Big|_{x=0} + \phi^2\Big|_{x=1} + \int_0^1 L\phi L\phi dx$$

$$= (1)^2 + (2)^2 + \int_0^1 (-2)(-2)dx = 9$$

Meanwhile,

$$\sum_{j=1}^{4} \gamma_j^{*2} = (3/\sqrt{2})^2 + (1/\sqrt{2})^2 + (-2)^2 + (0)^2 = 9 = (\phi,\phi).$$

Bessel's inequality can be used to evaluate the error of approximation for __Example 5.3.2__ as follows:

$$(\phi,\phi) = \int_{t=0}^{2} (L\phi)^2 dt = \int_{t=0}^{2} [q(t)]^2 dt = \int_0^1 (t^3)^2 dt + \int_1^2 (-2t^2 + 7t-4)^2 dt$$
$$= 3.6095$$

In comparison,

$$\sum_{j=1}^{3} \gamma_j^{*2} = (1.8575)^2 + (0.21082)^2 + (-0.325)^2 = 3.6003 \leq (\phi,\phi)$$

The generalized Fourier coefficients provide for the best approximation in the space S_m. The error of approximation X is given by

$$X = (\phi,\phi) - \sum_{j=1}^{m} \gamma_j^{*2} \qquad (5.4.3)$$

Because X is nonzero, the addition of independent basis elements to $<f_j>^m$ (increasing the dimension of S_m) will typically add more positive value to the sum of the γ_j^{*2}, resulting in a decrease in X. Should $X = 0$, then $||\phi - \hat{\phi}_m|| = 0$ and $(\phi - \hat{\phi}_m) = [0]$, the zero element, and $\phi = \hat{\phi}_m$ ae. For instance, example 5.3.1 results in $X = 0$, which indicates that the approximation $\hat{\phi}_4$ equals the exact solution ϕ ae (almost everywhere). Of course for this example, $\phi = \hat{\phi}_4$ identically over Ω, and the ae statement can be dropped.

However in Example 5.3.2, $X > 0$, indicating that ϕ is not in the subspace spanned by the considered basis functions.

Example 5.4.1

Consider the ℓ_2 best fit of $\lambda_1 v_1 + \lambda_2 v_2$ to b where, V is the linear space spanned by $\{v_1, v_2\}$, and

$$v_1 = \begin{Bmatrix} 1 \\ 0 \\ 1 \\ 0 \\ 1 \\ 0 \\ 1 \\ 0 \\ 1 \\ 0 \end{Bmatrix} ; \quad v_2 = \begin{Bmatrix} 1 \\ 2 \\ 1 \\ 1 \\ 1 \\ 0 \\ 0 \\ 0 \\ 0 \\ 0 \end{Bmatrix} ; \quad b = \begin{Bmatrix} 1 \\ 2 \\ 3 \\ 1 \\ 2 \\ 3 \\ 1 \\ 2 \\ 3 \\ 4 \end{Bmatrix}$$

Let $\{f_1,\ f_2\} = \{\mathbf{v_1},\ \mathbf{v_2}\}$ be the basis of linear space S_2. Using the dot product for (,) on S_2, the $\{f_i\}$ are orthonormalized as follows:

$$\hat{g}_1 = f_1$$

$$(\hat{g}_1,\hat{g}_1) = (\mathbf{v_1}\cdot\mathbf{v_1}) = 5$$

$$||\hat{g}_1|| = \sqrt{5}$$

$$g_1 = \hat{g}_1/||\hat{g}_1|| = (1,0,1,0,1,0,1,0,1,0)/\sqrt{5}$$

$$(f_2,g_1) = (\sqrt{5}/5)(3)$$

$$\hat{g}_2 = f_2 - (3\sqrt{5}/5)(1,0,1,0,1,0,1,0,1,0)/\sqrt{5}$$

$$= (0.4,2,0.4,1,0.4,0,-0.6,0,-0.6,0)$$

$$(\hat{g}_2,\hat{g}_2) = 31/5$$

$$||\hat{g}_2|| = \sqrt{31/5}$$

$$g_2 = \hat{g}_2/||\hat{g}_2|| = (2,10,2,5,2,0,-3,0,-3,0)/\sqrt{155}$$

The generalized Fourier coefficients are

$$\gamma_1^* = (\mathbf{b},g_1) = 10/\sqrt{5} = 2\sqrt{5}$$

$$\gamma_2^* = (\mathbf{b},g_2) = 5/\sqrt{31/5} = 25/\sqrt{155}$$

Now

$$g_1 = f_1/\sqrt{5} \Rightarrow \gamma_1^* g_1 = \gamma_1^* f_1/\sqrt{5} = 2f_1$$

$$g_1 = (-3/\sqrt{155})f_1 + (5/\sqrt{155})f_2$$

hence $\gamma_2^* g_2 = (25/\sqrt{155})\ [(-3/\sqrt{155})f_1 + (5/\sqrt{155})f_2]$

$$= (-15/31)f_1 + (25/31)f_2$$

Thus the best approximation in V of $\mathbf{b} \notin V$ is

$$\gamma_1^* g_1 + \gamma_2^* g_2 = 2f_1 - (15/31)f_1 + (25/31)f_2$$

$$= (47/31)f_1 + (25/31)f_2$$

and $(\lambda_1, \lambda_2) = (47/31, 25/31)$.

Example 5.4.2

Let vectors $\mathbf{F}_1$ and $\mathbf{F}_2$ be in $\mathbf{R}^n$ where $\mathbf{F}_1$ and $\mathbf{F}_2$ are composed of the values of continuous functions $f_1(x)$ and $f_2(x)$ at n equally spaced evaluation points defined on the domain $\Omega = [0, L]$ where, for example,

$$\mathbf{F}_1 = (f_1(\Delta), f_1(2\Delta), \cdots, f_1(n\Delta))$$

$$(5.4.4)$$

$$\mathbf{F}_2 = (f_2(\Delta), f_2(2\Delta), \cdots, f_2(n\Delta))$$

where $\Delta = L/n$. Then the L_2 inner product of (f,g) is

$$(f_1, f_2) = \int_0^L f_1 f_2 \, dx \cong \Delta \sum_{i=1}^n f_1(i\Delta) f_2(i\Delta)$$

$$= \Delta \mathbf{F}_1 \cdot \mathbf{F}_2$$

and as $n \to \infty$, $\Delta(\mathbf{F}_1 \cdot \mathbf{F}_2) \to (f_1, f_2)$.

Letting $\Omega = [0, 4]$, the functions $\{f_1, f_2, f_3\} = \{1, x, x^2\}$ form a basis of the linear space S_3. Let $\phi = x^2 - 1$, where $\phi \in S_3$. Vectors in $\mathbf{R}^5$ are defined to represent the f_i by (here we depart from (5.4.4) by including an evaluation point at $x = 0$)

$$f_1 \to \mathbf{F}_1 = (1,1,1,1,1)$$
$$f_2 \to \mathbf{F}_2 = (0,1,2,3,4)$$
$$f_3 \to \mathbf{F}_3 = (0,1,4,9,16)$$

where the evaluation points used in Ω are the x-coordinates $(0,1,2,3,4)$. The best appoximation $\hat{\phi}_3$ in S_3 of ϕ is determined as follows:

The $\{F_i\}$ are orthonormalized using the ℓ_2 norm into vectors $\{G_i\}$ as follows:

$\hat{G}_1 = F_1$

$(\hat{G}_1,\hat{G}_1) = \hat{G}_1 \cdot \hat{G}_1 = 5$

$||\hat{G}_1|| = \sqrt{5}$

$G_1 = \hat{G}_1/||\hat{G}_1|| = (1,1,1,1,1)/\sqrt{5}$

$(G_1,F_2) = 10/\sqrt{5}$

$\hat{G}_2 = F_2 - (G_1,F_2)G_1$

$\quad = (0,1,2,3,4) - (10/\sqrt{5})(1,1,1,1,1)/\sqrt{5}$

$\quad = (-2,-1,0,1,2)$

$(\hat{G}_2,\hat{G}_2) = 10$

$||\hat{G}_2|| = \sqrt{10}$

$G_2 = (-2,-1,0,1,2)/\sqrt{10}$

$(G_1,F_3) = 30/\sqrt{5}$

$(G_2,F_3) = 40/\sqrt{10}$

$\hat{G}_3 = F_3 - (G_1,F_3)G_1 - (G_2,F_3)G_2$

$\quad = (0,1,4,9,16) - (30/\sqrt{5})(1,1,1,1,1)/\sqrt{5}$

$\quad\quad\quad\quad\quad - (40/\sqrt{10})(-2,-1,0,1,2)/\sqrt{10}$

$\quad = (0,1,4,9,16) - (6,6,6,6,6) - (-8,-4,0,4,8)$

$\quad = (2,-1,-2,-1,2)$

$(\hat{G}_3,\hat{G}_3) = 14$

$||\hat{G}_3|| = \sqrt{14}$

$G_3 = (2,-1,-2,-1,2)/\sqrt{14}$

The orthonormal vectors for $\{F_i\}$ are

$\{G_1,G_2,G_3\} = \{(1,1,1,1,1)/\sqrt{5}, (-2,-1,0,1,2)/\sqrt{10}, (2,-1,-2,-1,2)/\sqrt{14}\}$

The function to be approximated is $\phi = x^2-1$, which in the vector form in $\mathbf{R}^5$ is

$\Phi = (-1,0,3,8,15)$

The generalized Fourier coefficients are

$\gamma_1^* = (G_1,\Phi) = 25/\sqrt{5}$

$\gamma_2^* = (G_2,\Phi) = 40/\sqrt{10}$

$\gamma_3^* = (G_3,\Phi) = 14/\sqrt{14}$

Thus the ℓ_2 best approximation $\hat{\phi}$ in S_3 is

$$\hat{\phi} = \gamma_1^* G_1 + \gamma_2^* G_2 + \gamma_3^* G_3$$

$$= (25/\sqrt{5})\, G_1 + (40/\sqrt{10})\, G_2 + (14/\sqrt{14})\, G_3$$

Resolving the G_i into F_i components (as each $G_i \in S_3$),

$G_1 = F_1/\sqrt{5}$

$G_2 = F_2/\sqrt{10} - F_1(2/\sqrt{10})$

$G_3 = F_3/\sqrt{14} - F_2(4/\sqrt{14}) + F_1(2/\sqrt{14})$

Therefore using the F_i,

$\hat{\phi} = (5F_1) + (4F_2 - 8F_1) + (F_3 - 4F_2 + 2F_1)$

$\quad = -F_1 + F_3 = \lambda_1 F_1 + \lambda_3 F_3$

indicating $(\lambda_1, \lambda_2, \lambda_3) = (-1,0,1)$

or $\hat{\phi} = x^2 - 1$, where $\hat{\phi} = \lambda_1 f_1 + \lambda_2 f_2 + \lambda_3 f_3$.

Example 5.4.3

Using the vector information of Example 5.4.2. find the best approximation in the linear space S_3 (spanned by $\{f_1, f_2, f_3\} = \{1, x, x^2\}$) of $\phi = x^3$ for $\Omega = [0,4]$.

Solution. Given the above $\{F_i\}$ and $\{G_i\}$, let ϕ be the vector composed of $\phi = x^3$ evaluated at the given evaluation points. Thus $\phi = (0,1,8,27,64)$

Then

$$\gamma_1^* = (G_1, \phi) = 100/\sqrt{5}$$
$$\gamma_2^* = (G_2, \phi) = 154/\sqrt{10}$$
$$\gamma_3^* = (G_3, \phi) = 84/\sqrt{14}$$

Resolving $\hat{\phi}$ into the F_i basis vectors,

$$\hat{\phi} = (20\, F_1) + (15.4\, F_2 - 30.8\, F_1)$$
$$+ (6\, F_3 - 24\, F_2 + 12\, F_1)$$
$$= 1.2\, F_1 - 8.6\, F_2 + 6\, F_3$$

giving $(\lambda_1, \lambda_2, \lambda_3) = (1.2, -8.6, 6)$ or $\hat{\phi} = 1.2 - 8.6x + 6x^2$.

As a comparison, use $\{1, x, x^2\}$ to approximate $\phi = x^3$ on $[0,4]$ in L_2.

Then

$$\hat{g}_1 = 1$$

$$(\hat{g}_1, \hat{g}_1) = \int_0^4 dx = 4; \quad ||\hat{g}_1|| = 2$$

$$g_1 = 1/2$$

$$\hat{g}_2 = f_2 - (g_1, f_2) \, g_1$$

$$(g_1, f_2) = 1/2 \int_0^4 x \, dx = 4$$

$$\therefore \hat{g}_2 = x - (4)(1/2) = x - 2$$

$$(\hat{g}_2, \hat{g}_2) = \int_0^4 (x-2)^2 \, dx$$

$$||\hat{g}_2|| = 4/\sqrt{3}$$

$$g_2 = \hat{g}_2 / ||\hat{g}_2|| = (x-2)/(4/\sqrt{3})$$

$$\hat{g}_3 = f_3 - (g_1, f_3) \, g_1 - (g_2, f_3) \, g_2$$

$$(g_1, f_3) = \int_0^4 (1/2) \, x^2 \, dx = 32/3$$

$$(g_2, f_3) = \int_0^4 ((x-2)/(4/\sqrt{3})) \, x^2 \, dx = 16/\sqrt{3}$$

$$\hat{g}_3 = x^2 - 4x + 8/3$$

$$(\hat{g}_3, \hat{g}_3) = \int_0^4 (x^2 - 4x + 8/3)^2 \, dx = 256/45$$

$$||\hat{g}_3|| = 16/3\sqrt{5}$$

$$g_3 = (x^2 - 4x + 8/3)(3\sqrt{5}/16)$$

$$\gamma_1^* = \int_0^4 g_1 \, x^3 \, dx = 32$$

$$\gamma_2^* = \int_0^4 g_2 \, x^3 \, dx = 96\sqrt{3}/5$$

$$\gamma_3^* = 32/\sqrt{5}$$

Back substituting, the best approximation is now

$$\hat{\phi}(x) = 6\,x^2 - 9.6\,x + 3.2$$

In comparison, the $\mathbf{R}^5$ vector representations gave

$$\hat{\phi}(x) = 6\,x^2 - 8.6\,x + 1.2$$

Example 5.4.4

An inner product with respect to the weighting function $W \geq 0$ is

$$(f_1, f_2) = \int_{\Omega} f_1 f_2 W d$$

Using vector representations F_i for functions f_i on Ω, a possible weighting is (recalling Example 5.4.2)

$$(f_1, f_2) \approx (\mathbf{F}_1, \mathbf{F}_2) \cdot \mathbf{W}$$

where $\mathbf{W} = (W(\Delta),\ W(2\Delta), \cdots, W(n\Delta))$.

Another weighting is the variation in density of evaluation points in Ω. For example, using $\Delta' = \Delta/2$ for region Ω' in Ω results in a doubling in the number of evaluation points in Ω', and hence a heavier weighting in Ω'.

5.5 The Weighted Inner Product

In the inner product of (5.2.2), equal weight is given to the various requirements imposed on the best approximation function ϕ_m from the space S_m spanned by the m linearly independent basis elements $<f_j>^m$. Namely, the L_2 error in satisfying the linear operator relationship over Ω is considered to be of equal importance as the L_2 error in satisfying the problem's boundary (and initial) conditions, (of course for the Voltera integral example problem, only one term is used in the inner product definition and the concerns as to inner product weighting factors is no longer needed).

Due to the limitations of computer power, only a finite number of basis functions can be used for approximation purposes. An argument is made to weight the terms which compose the inner product differently, in order to focus computational effort. For $0 < \varepsilon < 1$, one weighting of (5.2.2) is simply

$$(u,v) = \varepsilon \int_{\Gamma} uv d\Gamma + (1-\varepsilon) \int_{\Omega} LuLv d\Omega \qquad (5.5.1)$$

In (5.5.1), an ε-value close to 1 would force the approximation function $\hat{\phi}_m$ of S_m to focus upon satisfying the problem's boundary conditions rather than satisfying the operator in Ω. Similarly, the ε-value close to 0 would focus the $\hat{\phi}_m$ approximation towards satisfying the operator relationship in Ω and ignore the boundary conditions.

It is noted that (5.5.1) is still an inner product for any given choice of ε, and may be used to develop the generalized Fourier series using the previously presented procedures. And as the dimension S_m increases, the Bessel's inequality still applies except that now $X = X_\varepsilon$, and

$$X_\varepsilon = 0 \Rightarrow ||\hat{\phi}_m - \phi||_\varepsilon = 0 \tag{5.5.2}$$

In (5.5.2), the ε-notation has been added to the norm in order to clarify that all norms, inner products, and even the orthonormalized basis functions are now functions of ε. However for ease of presentation in the following text, the ε-notation is omitted although it is implied that all relationships are now dependent on the ε-value used in the weighting of the inner product components.

The selection of the "optimum" ε-value to be used in (5.5.1) depends on the rule assigned for optimization. One strategy is to choose ε which minimizes the Bessel's inequality relationship

$$X_\varepsilon = (\phi,\phi)_\varepsilon - \sum_{j=1}^{m} \gamma^{*2}_{\varepsilon_j} \tag{5.5.3}$$

$$= (\phi,\phi)_\varepsilon - \sum_{j=1}^{m} (\phi,g_{\varepsilon_j})^2_\varepsilon \tag{5.5.4}$$

In (5.5.3) and (5.5.4) it is stressed that all terms depend on ε.

Example 5.5.1

To illustrate the inner product weighting concept, Example 5.3.1, is restudied with only one basis function, $f_1 = x^2$. Recall that $L\phi = \dfrac{d^2\phi}{dx^2}$, h = -2, and $\phi(x=0) = 1$, $\phi(x=1) = 2$.

Proceeding as before, and dropping the ε subscript notation,

$$(f_1,f_1) = \varepsilon \int_\Gamma (f_1)^2 d\Gamma + (1-\varepsilon) \int_\Omega (Lf_1)^2 d\Omega$$

$$= \varepsilon(x^2)^2 \Big|_{x=0} + \varepsilon(x^2)^2 \Big|_{x=1} + (1-\varepsilon) \int_0^1 (2)^2 dx = 4-3\varepsilon$$

$$\therefore \|f_1\| = x^2/\sqrt{4-3\varepsilon}$$

and $g_1 = f_1/\|f_1\| = x^2/\sqrt{4-3\varepsilon}$

The only Fourier coefficient γ_1^* is computed as

$$\gamma_1^* = (\phi,g_1) = \varepsilon \left(\frac{x^2}{\sqrt{4-3\varepsilon}}\right)(1) \Big|_{x=0} + \varepsilon \left(\frac{x^2}{\sqrt{4-3\varepsilon}}\right)(2) \Big|_{x=1}$$

$$+ (1-\varepsilon) \int_0^1 \left(\frac{2}{\sqrt{4-3\varepsilon}}\right)(-2)dx = \frac{(6\varepsilon-4)}{\sqrt{4-3\varepsilon}}$$

Thus, $\hat{\phi}_1 = \gamma_1^* g_1$

$$= x^2 \left(\frac{6\varepsilon-4}{4-3\varepsilon}\right); \quad \text{for } 0 \le \varepsilon \le 1.$$

The next step is to compute χ_ε:

$$(\phi,\phi) = \varepsilon(\phi_b)^2\Big|_{x=0} + \varepsilon(\phi_b)^2\Big|_{x=1} + (1-\varepsilon)\int_0^1 h^2 dx$$

$$= \varepsilon(1)^2 + \varepsilon(2)^2 + (1-\varepsilon)\int_0^1 (-2)^2 dx$$

$$= 4+\varepsilon; \text{ for } 0 \le \varepsilon \le 1$$

$$\gamma_1^{*2} = (36\varepsilon^2 - 48\varepsilon + 16)/(4-3\varepsilon)$$

Therefore
$$\chi_\varepsilon = (\phi,\phi) - \gamma_1^{*2}$$
$$= (4+\varepsilon) - (36\varepsilon^2 - 48\varepsilon + 16)/(4-3\varepsilon)$$
$$= \varepsilon(40-39\varepsilon)/(4-3\varepsilon)$$

Figure 5.3 displays the plot of χ_ε against ε for $0 \le \varepsilon \le 1$. Because only one basis function $f_1 = x^2$ was chosen in this simple example, the weighting is focused on satisfying the operator or the boundary conditions as shown in Table 5.1. For this simple problem, $\hat{\phi}_1 = kx^2$ where $k = (6\varepsilon-4)/(4-3\varepsilon)$ from the above calculations. Table 5.1 summarizes the implications resulting from using values of k in $\hat{\phi}_1$. The maximum χ_ε value is computed by differentiation of the χ_ε formula above.

From Fig. 5.3 it is seen that χ_ε is minimum when $\varepsilon = 0$. Obviously from Table 5.1, however, $\varepsilon = 0$ would not be the optimum choice of ε due to the approximation only satisfying in a minimum least-squares sense the operator relationship in Ω and neglecting the boundary conditions, on Γ. One strategy is choosing ε that maximizes χ_ε. In this way, the "largest" value of approximation error is being used to evaluate Bessel's inequality, which is then used to evaluate the reduction in all possible weighted approximation errors (for the defined weighted norm) as additional elements are added to the collection of basis functions.

Table 5.1. Inner Product Weighting Implications for
Example 5.5.1

ϵ	χ_ϵ	k (for $\hat{\phi} = kx^2$)	Notes
0	0	-1.0	All weighting is focused toward satisfying $\frac{d^2\phi}{dx^2} = -2$. Here, $\hat{\phi}_1 = -x^2$.
0.50	4.1	-0.40	An intermediate approximation for $\hat{\phi}_1$
0.692	4.67	0.0790	Maximizes χ_ϵ
1.0	1.0	+2.0	All weighting is focused towards satisfying $\phi(x=0) = 1$ and $\phi(x=1) = 2$. Here $\hat{\phi}_1 = 2x^2$.

5.6. Considerations in Choosing Basis Functions

The previous example problems demonstrate that the approximation effort can only be as good as the set of basis functions used. Because we cannot use an infinite number of basis functions in a computer program, attempts to obtain exact solutions are typically impossible. However, some considerations are appropriate as to the choice of basis functions, in that some families of basis may be more successful in reducing approximation error (i.e., Bessel's inequality) than other families.

5.6.1. Global Basis Elements

Given the linear operator equation $L\phi = h$ on domain Ω, it is oftentimes useful to employ familiar functions such as multi-dimensional polynomials, trigonometric functions, and so forth. We will call these types of functions "global" functions in that these functions have nonzero value almost everywhere in Ω.

Example 5.6.1

Let the linear operator be given by the two-dimensional Laplace equation, $L\phi = 0$, where $\dfrac{\partial^2\phi}{\partial x^2}+\dfrac{\partial^2\phi}{\partial y^2} = 0$ in Ω with boundary conditions $\phi = \phi_b$ on the boundary Γ.

Some choices for global basis functions are $\{f_j\}= \{\sin x, \cos y, 1, xy, x^2, y^2\}$. This is a set of six linearly independent functions, and the set $\{Lf_j\} = \{-\sin x, -\cos y, 0, 0, 2, 2\}$. The best approximation in S_6 in solving our linear operator equation has the form $\hat{f}(x) = C_1\sin x + C_2\cos y + C_3 + C_4xy + C_5x^2 + C_6y^2$. Note that the two basis functions $\{1, xy\}$ have no influence in satisfying the Laplace equation operator, as $L(1) = 0$ and $L(xy) = 0$; however, these two functions could provide considerable help in approximating the boundary conditions.

Another choice for basis functions is $\{f_j\} = \{1, x, y, xy, (x^2 - y^2), (x^3 -3xy^2)\}$. Note that $\{Lf_j\} = \{0, 0, 0, 0, 0, 0\}$. In this family, we have the same dimension as the preceding family, yet any linear combination in the second family results in an approximation that exactly solves the operator in Ω. Thus, there is only error in approximating the boundary conditions. Our best approximation has the form $\hat{f}(x) = C_1 + C_2x + C_3y + C_4xy + C_5 (x^2-y^2) + C_6 (x^3-3xy^2)$. Hence, the best approximation is obtained by finding the C_i that minimizes the integral

$$\chi = \int_{\Gamma} (\phi_b(x,y)-C_1-C_2x-C_3y-C_4xy-C_5(x^2-y^2)-C_6(x^3-3xy^2))^2 \, d\Gamma$$

5.6.2. Spline Basis Functions

A convenient approach to developing approximation functions is to use spline functions that are zero everywhere except over a small region, or finite element, in the problem domain . This technique is very amenable to computer methods due to the respective calculations involved. Generally, piecewise continuous polynomial functions are involved which are defined to be zero outside of the specific finite elements assigned.

Example 5.6.2 (Spline Trial Functions)

For the same problem as Example 5.6.1, let $\Omega = \{(x,y): 0 \le x \le 2,$ $0 \le y \le 1\}$. The Laplace equation is $L\phi = 0$ on Ω. On the boundary, Γ, $\phi_b =$ 100 at $x = 0$; $\phi_b = 0$ at $x = 2$; and ϕ_b is linear with respect to x. Two triangular elements Ω^1 and Ω^2 are used that share a diagonal of the rectangular domain.

Each finite element contains six nodal points, ϕ_j (where ϕ_j is the estimate of ϕ at node j). In this example, spline functions $N_j(x,y)$ are defined as a set of polynomials of the form

$$N_j(x,y) = (C_0 + C_1 x + C_2 y + C_3 xy + C_4 x^2 + C_5 y^2)_j$$

where j is the node number, and the coefficients $(C_0, C_1, C_2, \cdots, C_5)_j$ are determined by setting

$$N_j(p_k) = \begin{cases} 1 ; & k=j \\ \\ 0 ; & k \ne j \end{cases}$$

where p_k is node k. The approximation function $N_j(x,y) = 0$ for all (x,y) exterior of the considered finite element. Usually the finite element number is also designated, such as N_5^1 or N_5^2 where N_5^1 is the node 5 spline function for element 1 whereas N_5^2 is the node 5 spline function for element 2. The global approximation function is written as (for the nine nodal points)

$$\Phi_9(x,y) = \phi_1 N_1^1 + \phi_2 N_2^1 + \phi_3 N_3^1 + \phi_4 N_4^1$$

$$+ \phi_5 N_5^1 + \phi_7 N_7^1 + \phi_3 N_3^2 + \phi_5 N_5^2 + \phi_6 N_6^2$$

$$+ \phi_7 N_7^2 + \phi_8 N_8^2 + \phi_9 N_9^2$$

or in summed form,

$$\Phi_9(x,y) = \sum_{j=1}^{9} \phi_j N_j$$

where, for example, $N_5 = N_5^1 + N_5^2$, when appropriate. The ϕ_j values are the unknowns to be solved directly using generalized Fourier series. That is, find the best approximation $\hat{\Phi}_9$ by choosing the optimum values of $\{\phi_1, \phi_2, \cdots, \phi_9\}$ that minimize $||\hat{\Phi}_9 - \phi||$ where

$$(u,v) = \int_\Omega LuLvd\Omega + \int_\Gamma uvd\Gamma$$

The evaluation point coordinates used for the vector representations are contained in the vector **C**, where **C** is a vector of geometric coordinates of evaluation points used in numerical integration, and where auxilliary conditions are evaluated. There are eight elevation points defined on Γ to use in minimizing the ℓ_2 error of $||\hat{\Phi}_b - \phi_b||$ on Γ, and two evaluation points in Ω (one each in Ω^1 and Ω^2) for minimizing ℓ_2 error of $||L\hat{\Phi}_9|| = 0$ in Ω.

The coordinates for the evaluation points are ordered as follows:

C = ((0,.75), (0,.25), (.5,0), (1.5,0), (2,.25), (2,.75),

(1.5,1), (.5,1), (.5,.75), (1.5,.25))

where the first eight evaluation points are on the boundary, and the last two are inside Ω (and are used for L purposes in the numerical evaluation of the operator, in vector form).

Consider first finite element Ω^1.

$P_1^1 = \frac{1}{2}x - 1y - 2xy + \frac{1}{2}x^2 + 2y^2$

$LP_1^1 \ k = 5$

To find $\mathbf{F}_1$, substitute each of the evaluation points along the boundary of Ω_1 into P_1. The contribution of $\mathbf{F}_1$ at the evaluation point within Ω^1 is $LP_1^1 = 5$, whereas in Ω^2, $LP_1^1 = 0$.

$\mathbf{F}_1 = (.375, -.125, .375, 1.875, 0, 0, 0, 0, 5, 0)$

Now $P_2^1 = -2x + 4y + 2xy - 4y^2$

$LP_2^1 = -8$

$\mathbf{F}_2 = (.75, .75, -1, -3, 0, 0, 0, 0, -8, 0)$

$P_3^1 = 1 - 3y + 2y^2$

$Lp_3^1 = 4$

$\mathbf{F}_3 = (-.125, .375, 1, 1, 0, 0, 0, 0, 4, 0)$

$P_5^1 = 2x - 2xy$

$LP_5^1 = 0$

$\mathbf{F}_5 = (0, 0, 1, 3, 0, 0, 0, 0, 0, 0)$

$P_7^1 = -\frac{1}{2}x + \frac{1}{2}x^2$

$LP_7^1 = 1$

$\mathbf{F}_7 = (0, 0, -.125, .375, 0, 0, 0, 0, 1, 0)$

$P_4^1 = 2xy - x^2$

$LP_4^1 = -2$

$\mathbf{F}_4 = (0, 0, -.25, -2.25, 0, 0, 0, 0, -2, 0)$

Now consider Ω^2:

$P_3^2 = 1 - \frac{3}{2}x + \frac{1}{2}x^2$

$LP_3^2 = 1$ Adding this element's contribution to F_3 yields (recalling that $N_3 = N_3^1 + N_3^2$):

$F_3 = (-.125,.375,1,1,0,0,-.125,.375,4,1)$

$P_6^2 = 2x - 4y + 2xy - x^2$

$LP_6^2 = -2$

$F_6 = (0,0,0,0,0,0,-.25,-2.25,0,-2)$

$P_9^2 = -\frac{1}{2}x + y - 2xy + \frac{1}{2}x^2 + 2y^2$

$LP_9^2 = 5$

$F_9 = (0,0,0,0,.375,-.125,.375,1.875,0,5)$

$P_8^2 = 2xy - 4y^2$

$Lp_8^2 = -8$

$F_8 = (0,0,0,0,.75,.75,-1,-3,0,-8)$

$P_7^2 = -y + 2y^2$

$LP_7^2 = 4.$ Adding this element's contribution to F_7 yields:

$F_7 = (0,0,-.125,.375,-.125,.375,1,1,1,4)$

$P_5^2 = 4y-2xy$

$LP_5^2 = 0.$ Adding this element's contribution to F_5 yields:

$F_5 = (0,0,1,3,0,0,1,3,0,0)$

Let $\Phi = (100,100,75,25,0,0,25,75,0,0)$,

where ϕ_b is known on Γ and $L\phi = 0$ in Ω.

The final vector representations $\{F_j\}$ of the basis $\{N_j\}$ are

$F_1 = (.375,-.125,.375,1.875,0,0,0,0,5,0)$

$F_2 = (.75,.75,-1,-3,0,0,0,0,-8,0)$

$F_3 = (-.125,.375,1,1,0,0,-.125,.375,4,1)$

$F_4 = (0,0,-.25,-2.25,0,0,0,0,-2,0)$

$F_5 = (0,0,1,3,0,0,1,3,0,0)$

$F_6 = (0,0,0,0,0,0,-.25,-2.25,0,-2)$

$F_7 = (0,0,-.125,.375,-.125,.375,1,1,1,4)$

$F_8 = (0,0,0,0,.75,.75,-1,-3,0,-8)$

$F_9 = (0,0,0,0,.375,-125,.375,1.875,0,5)$

and $\Phi = (100,100,75,25,0,0,25,75,0,0)$.

The best approximation of $\Phi_9 = \sum_{i=1}^{9} \lambda_i F_i$ is determined by computing the generalized Fourier series coefficients $\{\lambda_i\}$. Note that in using the $\{N_j\}$ spline functions, the $\lambda_i = \phi_i$, i=1,2,$\cdots$,9. That is,

$$\Phi_9 = \sum_{i=1}^{9} \phi_i \ N_i(x,y).$$

5.6.3. Mixed Basis Functions

The use of spline and global basis functions together in a set of approximation elements is handled no differently. The basis dimension is equal to the number of linearly independent elements. Of value in the best approximation method is the ability to add particular global functions that are singular in the problem domain (e.g., $\frac{1}{x}, \frac{1}{x^2}$, etc.) or have special properties, to the set of basis functions. In chapter 6, example problems demonstrate the freedom available to this numerical method's choice of basis elements.

CHAPTER 6
THE BEST APPROXIMATION METHOD: APPLICATIONS

6.0. Introduction

The theory of generalized Fourier Series as utilized in the Best Approximation Method can be applied to the approximation of linear operator relationships. To demonstrate the computational results in using this approach, several example problems where analytic solutions or quasi-analytic solutions exist are considered. Applications include two-dimensional problems involving the Laplace and Poisson equations, tests for variation in results due to inner product weighting factors, and applications to nonhomogeneous domain problems.

This chapter focuses on the topics of weighting factor selection and modeling sensitivity, effects on additional basis functions on computational accuracy, and the effects on modeling results due to the addition of collocation points. Thirteen simple but detailed example problems are used to illustrate the approximate results obtained by the method when applied to practical problems involving Partial Differential Equations (PDE) in transport type problems. Chapter 7 focuses upon use of analytic functions for approximating two-dimensional transport problems involving the Laplace or Poisson equations, demonstrating how the choice of basis functions influences the computational effort. A FORTRAN computer program for the Best Approximation Method is included as a final section of this chapter.

6.1. Sensitivity of Computational Results to Variation in the Inner Product Weighting Factor

The inner product uses a weighting factor, ε, to weight the approximation effort in satisfying the PDE and boundary condition (BC) values by

$$(u \cdot v) = \varepsilon \int_\Gamma u \cdot v \, d\Gamma + (1-\varepsilon) \int_\Omega Lu \, Lv \, d\Omega$$

The effects of varying ε between 0 and 1 is shown in the following simple applications.

Example 6.1.1 (Weighting Factor Influence)

Let $\phi = 2x^2y + y^2 + x + 6$ where $\nabla^2\phi = 2 + 4y$ on the unit square domain. Figure 6.1 depicts the problem domain and boundary conditions.

First, define $<1,x,y,xy,x^2>$ to be the set of basis functions, and use $\varepsilon = 0$. The resulting approximation function is $\hat{\phi} = 2x^2$.

Applying the linear operator ∇^2 on $\hat{\phi}$, we obtain $\nabla^2\hat{\phi} = 4$. The graph of $\nabla^2\phi = 2 + 4y$ on the unit square domain is depicted in Fig. 6.2. Note that $\{Lf_i\} = <0,0,0,0,2>$. Thus when $\varepsilon = 0$, the approximation effort becomes finding the best fit of $\lambda_5(2)$ to $2 + 4y$ in Ω. In the computer program application, the vector of coordinates used is

$$C = [(0.5,0),(1,0.5), (0.5,1), (0,0.5), (0.5,0.5)]$$

where boundary conditions are being considered with four evaluation points, and a single evaluation point is used for considering the operator. The $\{f_i\}$ vectors are

$F_1 = (1, 1, 1, 1, 0)$

$F_2 = (0.5, 1, 0.5, 0, 0)$

$F_3 = (0, 0.5, 1, 0.5, 0)$

$F_4 = (0, 0.5, 0.5, 0, 0)$

$F_5 = (0.25, 1, 0.25, 0, 2)$

and $\Phi = (6.5, 8.25, 8, 6.25, 4)$

For $\varepsilon = 0$, the $\{F_i\}$ become zero vectors (and hence are neglected) except for $F_5 = (0,0,0,0,2)$. And, the best approximation is $\hat{\phi}_5 = 2x^2$. From Fig. 6.2, it is seen that indeed for $\varepsilon = 0$, the best approximation for $L\phi = h$ is $\nabla^2\hat{\phi} = 4$. Thusly, the $\nabla^2\hat{\phi}$ approximation satisfies the linear operator on the least square sense (ℓ_2) for $\varepsilon = 0$ for the given set of basis functions.

Second, we choose $<1,x,y>$ as the set of basis functions and use $\varepsilon = 1$. Then the resulting approximation function is $\hat{\phi} = 5.4583 + 2x + 1.75y$. For comparison, the least squares method can be used to minimize the residual

$$X = \int_{\Gamma} (k_1 + k_2x + k_3y - \phi_b)^2 \, d\Gamma$$

with respect to parameters k_1, k_2, and k_3 along the boundary of the unit square domain. From Fig. 6.1, one can write X as follows:

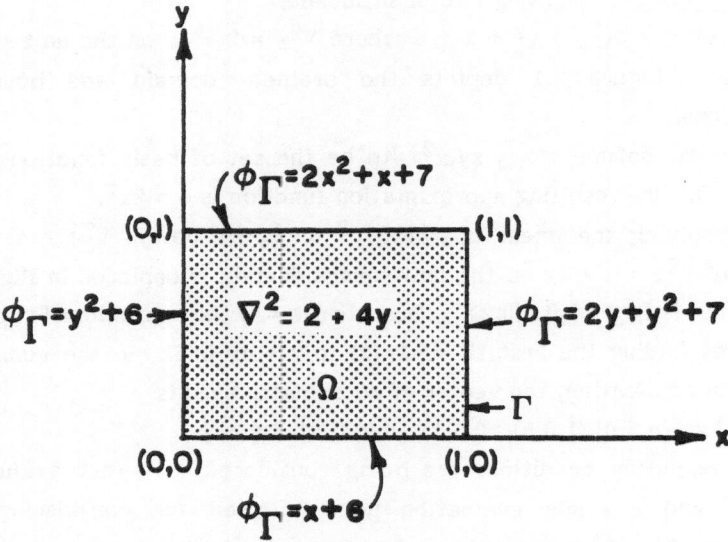

Fig. 6.1. Domain and Boundary Conditions for Example 6.1.1.

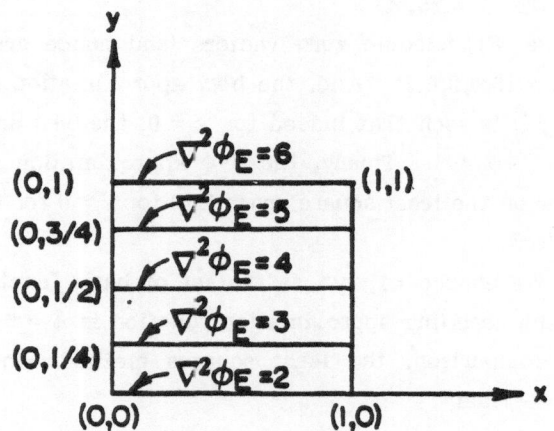

Fig. 6.2. Graphs for Exact Solution of $\nabla^2 \phi = 2 + 4y$ on Unit Square Domain.

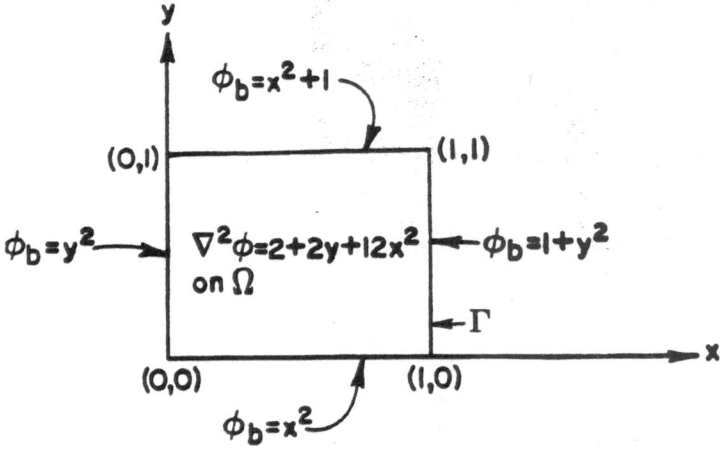

Fig. 6.3.   Domain and Boundary Conditions for Example 6.1.2.

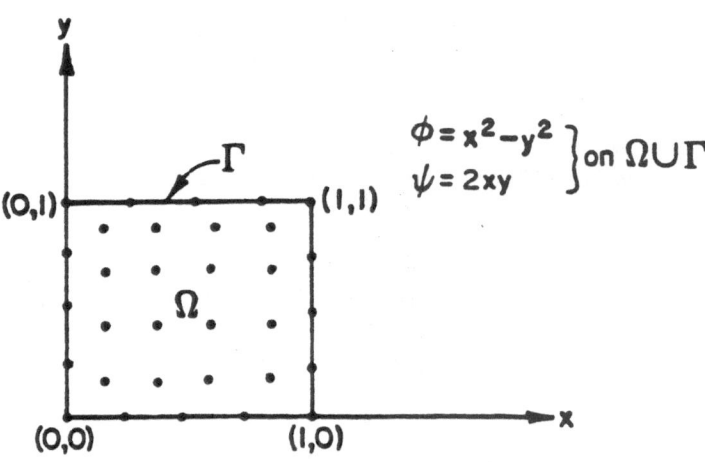

Fig. 6.4.   Domain and Nodal Placement for Example 6.2.1.

$$\chi = \int_{x=0}^{1} (k_1 + k_2 x - x - 6)^2 dx + \int_{y=0}^{1} (k_1 + k_2 + k_3 y - 2y - y^2 - 7)^2 dy$$

$$+ \int_{x=0}^{1} (k_1 + k_2 x + k_3 - 2x^2 - x - 7)^2 dx + \int_{y=0}^{1} (k_1 + k_3 y - y^2 - 6)^2 dy$$

Minimizing χ with respect to k_1, k_2, and k_3, we obtain

$$k_1 = 5.4583$$
$$k_2 = 2.0$$
$$k_3 = 1.75$$

which verifies that the approximation function is the least square fit (ℓ_2) with respect to the problem boundary conditions.

Example 6.1.2 (Weighting Factor Sensitivity)

Let's consider a Poisson problem $\nabla^2 \phi = 2 + 2y + 12x^2$ on an unit square domain with boundary condition $\phi_b = x^2 + y^2$. (Fig. 6.3)

By choosing the following set of basis functions $<x^2, x^3, x^4, y^2, y^3, y^4, xy^2, yx^2, x^2 y^2>$, we obtain the approximation function $\hat{\phi} = x^2 + y^2$ for $\epsilon = 1$ and $\hat{\phi} = x^2 + \frac{1}{3} y^3 + x^4$ for $\epsilon = 0$ with $\nabla^2 \hat{\phi} = 2 + 2y + 12x^2$. For $\epsilon = 0$ and 1, we obtain the two extreme values of the weighting factor $\epsilon = 0$ and 1 in satisfying the PDE on the domain and satisfying the BC values on the boundary, respectively.

6.2. Solving Two-Dimensional Potential Problems

Presented in the following are application problems and computed results in solving the Laplace equation in two-dimensions. Because the Laplace equation can be phrased as a linear operator problem, the Best Approximation Method approach is appropriate.

Example 6.2.1 (Ideal Fluid Flow Around a 90° Bend)

The flow of an ideal fluid around a 90° bend can be expressed by the analytic function

$$\omega = z^2$$
$$= (x^2 - y^2) + 2xyi$$

Since the state variable function $\phi = (x^2 - y^2)$ and the stream function $\psi = (2xy)$ are of polynomial form, the approximation basis $\{f_i\} = \{1, x, y, x^2, xy, y^2\}$ functions for the state variable and stream functions are found to result in the exact solutions regardless of the value of the weighting factor ε, $(0 < \varepsilon < 1)$. Figure 6.4 depicts the problem domain and nodal point placement used for this application. The **C** vector is composed of 16 boundary evaluation points and 16 domain evaluation points for use in minimizing the ℓ_2 error norm $||\hat{\phi}_b - \phi_b|| + ||L\hat{\phi} - L\phi||$.

Example 6.2.2 (St. Venant Torsion)

Consider the St. Venant torsion problem for an equilateral triangular section of Fig. 6.5. The analytic solution for $\phi(x,y)$ is given by

$$\phi(x,y) = (x^3 - 3xy^2)/2a + 2a^2/27$$

Figure 6.5 depicts the evaluation point placement used in the vector representations of the basis functions, and boundary conditions for the case of $a = 3$. The ℓ_2 approximation function $\hat{\phi}_8$ results in the exact solution when the basis $\{f_i\} = \{1, x, y, x^2, xy, y^2, x^3, y^3\}$ is used, due to $\phi \in S_8$. The evaluation point **C** vector contained 12 boundary and 15 domain points.

Example 6.2.3 (Potential Problem: Sharp Corner Domain)

Figure 6.6 depicts the evaluation point placement and boundary conditions for a mechanical gear problem. The resulting ℓ_2 approximation function is $\hat{\phi} = 9.518 + .3131x + .1812y + .7686xy + .2219x^2 - .2228y^2 - 1.018x^2y + .3397y^3 - .2463x^3y + .2124x^2y^2 + .2419xy^3$ for a weighting factor of $\varepsilon = 0.5$. Table 6.1 compares the results from a Complex Variable Boundary Element Method or CVBEM (see Chapter 7) model and the ℓ_2 approximation function for the interior nodal points, and the defined operator relationship. The **C** vector used contained 21 boundary and 21 interior domain evaluation points for use in minimizing boundary value and operator error norms, respectively.

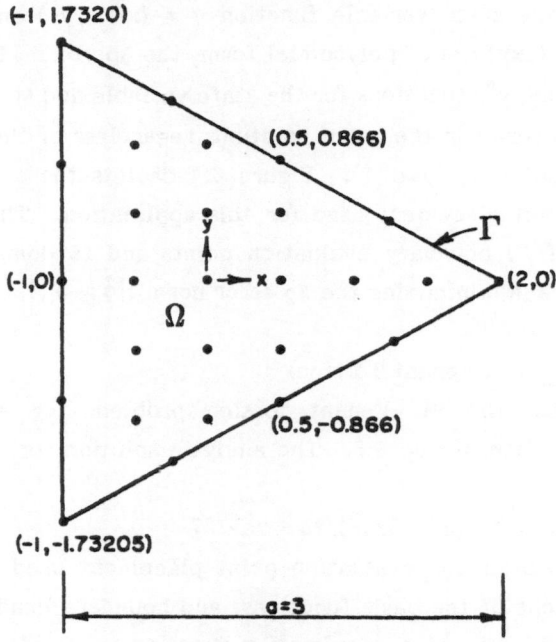

Fig. 6.5. Domain and Nodal Placement for Example 6.2.2.

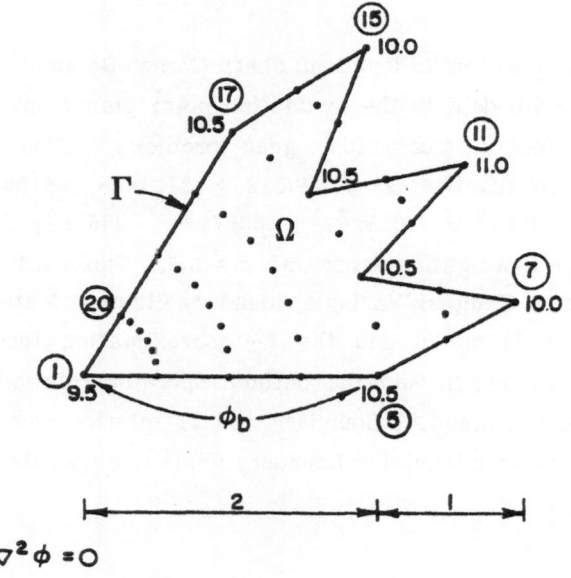

$$\nabla^2 \phi = 0$$

Fig. 6.6. Domain and Boundary Conditions for Example 6.2.3.

Table 6.1. Comparison of Best Approximation and CVBEM
Computational Results for <u>Example 6.2.3</u>

Interior Evaluation Coordinate Points

x	y	CVBEM	$\hat{\phi}$	$\nabla^2 \hat{\phi}$
.321	.383	9.7842	9.752	8.03×10^{-5}
.643	.766	10.0391	10.04	8.56×10^{-4}
.964	1.149	10.2610	10.28	6.36×10^{-4}
1.286	1.532	10.3668	10.39	-5.79×10^{-4}
1.607	1.915	10.2417	10.31	-2.79×10^{-3}
.492	.087	9.857	9.752	2.35×10^{-4}
.985	.174	10.0992	10.04	1.19×10^{-3}
1.477	.261	10.2749	10.28	1.18×10^{-3}
1.970	.347	10.3659	10.39	-2.98×10^{-3}
2.462	.434	10.2439	10.31	-1.76×10^{-3}
.470	.171	9.8476	9.767	-3.55×10^{-4}
.940	.342	10.0944	10.05	-1.32×10^{-3}
1.410	.513	10.2830	10.27	-4.50×10^{-3}
.433	.250	9.8329	9.772	-5.88×10^{-4}
.866	.50	10.0839	10.05	-2.26×10^{-3}
1.299	.75	10.2856	10.27	-1.67×10^{-3}
1.732	1.0	10.4917	10.42	-1.39×10^{-2}
2.165	1.25	10.7489	10.63	-2.39×10^{-2}
.383	.321	9.8119	9.767	-4.34×10^{-4}
.766	.643	10.0659	10.05	-1.49×10^{-3}
1.149	.964	10.2806	10.27	-4.87×10^{-3}

<u>Example 6.2.4</u> (Ideal Fluid Flow Around a Cylinder)

Ideal fluid flow around a cylinder has the analytic function model of (see Fig. 6.7 for problem domain shape)

$$\omega(x) = z + \frac{1}{z}$$

The state variable (potential) function can be expressed as

$$\phi(x,y) = x \left[1 + \frac{1}{x^2 + y^2} \right]$$

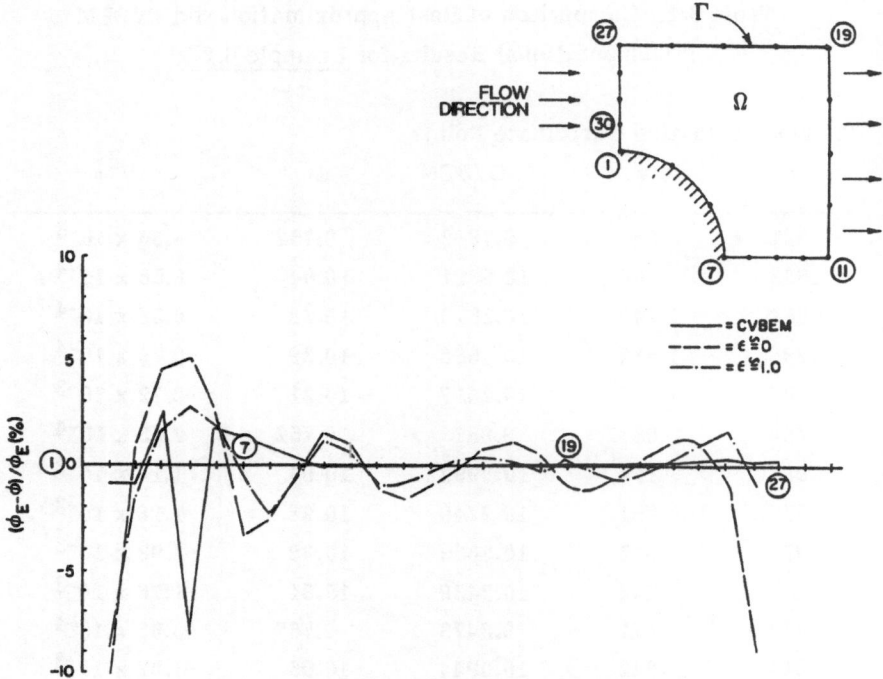

Fig. 6.7. Domain, Nodal Placement and Relative Error for Example 6.2.4.

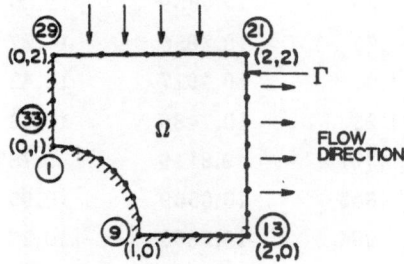

Fig. 6.8. Domain and Boundary Conditions for Example 6.2.5.

and the stream function can be expressed as $\psi(x,y) = y \left[1 - \dfrac{1}{x^2 + y^2} \right]$
where $\omega(z) = \phi(x,y) + i\psi(x,y)$. The approximations for ($\varepsilon = 0.5$) are

$$\hat{\phi} = 1.704 + .9795x - 2.587y + + 2.115xy - 1.009x^2 + 1.003y^2$$
$$- 1.289xy^2 + .431x^3 + .1452x^3y + .2258x^2y^2 - .1461xy^3$$

and

$$\hat{\psi} = -1.704 + 2.587x + 1.02y - 2.115xy - 1.003x^2 + 1.009y^2$$
$$+ 1.289x^2y - .4309y^3 + .1461x^3y - .2257x^2y^2 - 1.452xy^3$$

for state variable and stream functions, respectively. Figure 6.7 shows the approximation relative error between the CVBEM model and the approximation function values, for ε close to zero and one.

Example 6.2.5 (Ideal Fluid Flow Around a Cylinder in 90° Bend)

Ideal fluid flow around a cylinder in a 90° bend has the analytic solution of $\omega(z) = z^2 + z^{-2}$. The state variable function ϕ and stream function ψ can be expressed, respectively, as

$$\phi = (x^2 - y^2) \left[1 + \frac{1}{(x^2 + y^2)^2} \right]$$

and

$$\psi = 2xy \left[1 - \frac{1}{(x^2 + y^2)^2} \right]$$

The approximation functions ($\varepsilon = 0.5$) for the state variable function $\hat{\phi}$ and the stream function $\hat{\psi}$ are (for the polynomial basis functions $\{1,x,y,xy,x^2,y^2,x^2y,xy^2,x^3,y^3,x^3y,xy^3\}$)

$$\hat{\phi} = 3.197x - 3.197y - 1.828x^2 + 1.828y^2 + 1.894x^2y - 1.894xy^2$$
$$+ .6294x^3 - .6294y^3 + .4612x^3y - .4612xy^3$$

and

$$\hat{\psi} = -3.192 + 3.389x + 3.389y - 4.311xy + 1.352x^2 + 1.352y^2 + 2.121x^2y$$
$$+ 2.121xy^2 - .7136x^3 - .7136y^3 - .7536x^2y^2 + .1259x^4 + .1259y^4$$

on the domain shown in Fig. 6.8. In comparison of the approximation to exact values it was found that the relative error is high along the circumference of the cylinder.

If we included some singular elements, e.g., $\frac{1}{x}, \frac{1}{y}, \frac{1}{x^2}, \frac{1}{y^2}, \cdots$, into the set of basis functions $\{f_i\}$, the approximation function for the state variable $\hat{\phi}$ becomes

$$\hat{\phi} = 1.892x - 1.892y - 1.432x^2 + 1.432y^2 + .4409x^2y$$
$$- .4409xy^2 - \frac{.434}{x} + \frac{.434}{y} + \frac{.1237}{x^2} - \frac{.1237}{y^2}$$
$$- \frac{.01721}{x^3} + \frac{.01721}{y^3}$$

By increasing the boundary evaluation points from 32 to 60 and the number of basis functions, we obtain the approximation function

$$\hat{\phi} = 3.324x - 3.324y - 2.162x^2 + 2.162y^2 + .3721x^2y$$
$$- .3721xy^2 + .6896x^3 - .6896y^3 - \frac{2.336}{x} + \frac{2.336}{y}$$
$$+ \frac{.9113}{x^2} - \frac{.9113}{y^2} - \frac{.1609}{x^3} + \frac{.1609}{y^3} + \frac{.01315}{x^4} - \frac{.01315}{y^4}$$

Example 6.2.6 (Flow Net for Soil-Water Flow)

Figure 6.9 depicts the flow net for soil-water flow through a homogeneous soil as computed by the CVBEM. The approximation function of the state variable function $\hat{\phi}$ is given by

$$\hat{\phi} = 24. - .6549x - .0989y - 1.864xy - .06852x^2$$
$$+ 0.118y^2 + .0024x^2y + .001676xy^2 + .001878x^3 - .0003129y^3$$

for $\varepsilon = 0.5$. Table 6.2 shows the interior nodal point values for both the CVBEM and the ℓ_2 approximation function.

Note that in all the above applications, polynomial or rational polynomial basis functions are used. Also, vector representations $\{\mathbf{F}_i\}$ are used instead of the $\{f_i\}$ basis elements, with the usual dot product being the inner product, and ε being a weighting factor used to offset dimensionality inequaties due to $\Delta\Omega$ and $\Delta\Gamma$ differences (as seen in Example 5.4.5).

145

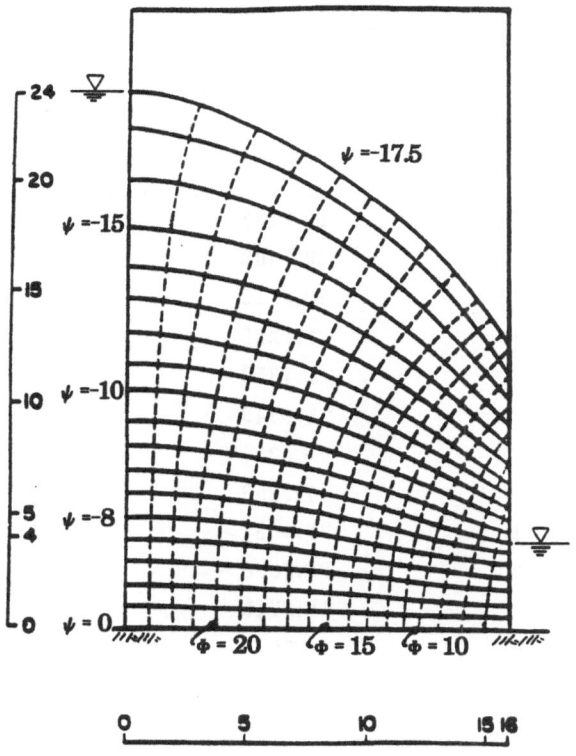

Fig. 6.9.   Streamlines and Potentials for Soil-Water Flow
Through a Homogeneous Soil for Example 6.2.6.

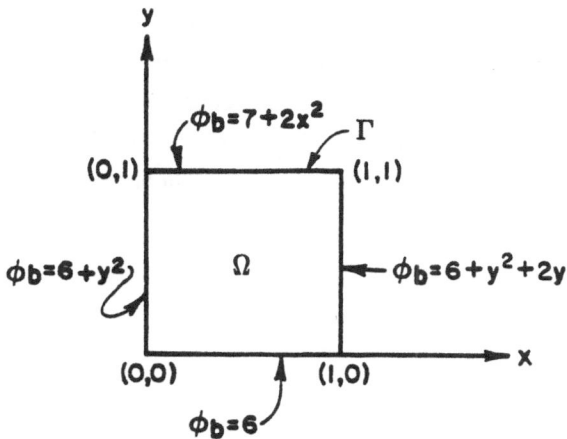

Fig. 6.10. Domain and Boundary Conditions for Example 6.3.1.

Table 6.2. CVBEM and Best Approximation Method
Results for Example 6.2.6

x	y	CVBEM	$\hat{\phi}$
4	4	19.9233	2014
4	8	20.3408	20.35
4	12	20.9110	20.91
4	16	21.5258	21.70
4	20	22.1359	22.61
8	4	15.6543	15.34
8	8	16.5817	16.04
8	12	17.8253	17.30
8	16	19.0937	19.01
8	20	20.2656	21.04
12	4	10.7715	10.11
12	8	12.5019	11.6
12	12	14.7428	13.87
12	16	16.8199	16.8

6.3. Application to Other Linear Operators

The best approximation method can be applied to other linear operators. The example problems of Chapter 5 considered a linear integral equation, and two differential equation problems. This section focuses upon application of the best approximation method to sophisticated problems involving potential theory.

Example 6.3.1 (Potential Flow in an Anisotropic Domain)

Consider the linear operator, $\nabla^2\phi$,

$$\nabla^2\phi = T_{xx}\frac{\partial^2\phi}{\partial x^2} + T_{yy}\frac{\partial^2\phi}{\partial y^2} \text{ on the unit square domain with}$$

boundary condition $\phi_b = 6 + y^2 + 2x^2y$ (Fig. 6.10).

The approximation functions for different T_{xx} and T_{yy} values are shown in Table 6.3. Included in Table 6.3 are the generalized Fourier series coefficients (after resolution through the back substitution process) for the considered basis functions (as tabulated).

Table 6.3. Best Approximation Function Coefficients
and Relative Errors for Example 6.3.1

T_{xx}	T_{yy}	1	x	y	xy	x^2	y^2	x^2y	xy^2	x^3	y^3	Maximum Relative error on Boundary	Maximum Relative error on Operator Relationship
1	1	5.856	.8996	.6616	.2092	-.8946	.8959	1.791	= 0	= 0	-.5967	10^{-2}	10^{-2}
2	1	5.878	.8277	.3158	.4275	-.8277	1.656	1.572	= 0	= 0	-1.048	10^{-2}	10^{-3}
1	2	5.849	.8657	.8637	.1007	-.8657	.4333	1.899	= 0	= 0	-.3165	10^{-2}	10^{-3}

Example 6.3.2 (Poisson Problem)

Consider $\phi = 6 + y^2 + 2x^2$ on the unit square domain (first quadrant) with the linear operator $\nabla^2\phi = \frac{\partial^2\phi}{\partial x^2} + \frac{\partial^2\phi}{\partial y^2}$, such that

$\nabla^2\phi = 2 + 4y$ in Ω.

The best approximation functions are listed in Table 6.4 for the various values of the weighting factor, ε.

Table 6.4. L_2 Solution of Poisson Problem for

<u>Example 6.3.2</u>

ε	$\hat{\phi}$
0	$x^2 + 2x^2y$ cr $y^2 + 2x^2y$
$0 < \varepsilon < 1$	$6 + y^2 + 2x^2y$
$\varepsilon = 1$	$6 + y^2 + 2x^2y$

<u>Example 6.3.3</u> (Poisson Problem in Anisotropic Domain)

Consider the linear operator, $\nabla^2\phi = T_{xx} \dfrac{\partial^2\phi}{\partial x^2} + T_{yy} \dfrac{\partial^2\phi}{\partial y^2}$, on the unit square domain. Then

$$\nabla^2\phi = 4\,T_{xx} + 2yT_{yy}$$

The best approximation functions for different values of T_{xx} and T_{yy} are listed in Table 6.5 for $\varepsilon = 0.5$.

Table 6.5. Best Approximation Results for

<u>Example 6.3.3</u>

T_{xx}	T_{yy}	$\hat{\phi}$
1	1	$6 + y^2 + 2x^2y$
2	1	$6.014 - .085x - .305y + .171xy + .085x^2$
		$+ 1.829y^2 + 1.829x^2y - .553y^3$
1	2	$5.993 + .0337x + .174y - .0675xy$
		$- .0337x^2 + .517y^2 + 2.067x^2y + .3221y^3$
2	2	$6 + y^2 + 2x^2y$

The maximum relative error on the boundary is of magnitude 10^{-4} and on the interior is of magnitude 10^{-5} for $T_{xx} \neq T_{yy}$.

<u>Example 6.3.4</u> (Poisson Problem in Triangular Domain)

Consider <u>Example 6.3.2</u> for the problem domain of <u>Example 6.2.2</u> (Fig. 6.5). The approximation function provides the exact solution regardless of the value of the weighting factor ($0 < \varepsilon < 1$).

<u>Example 6.3.5</u> (Poisson Problem with Transcendental Functions)

Consider $\phi = \sin x + yx + 10 + \cos^2 y$ on a unit square domain with a linear operator $\nabla^2\phi$, such that $\nabla^2\phi = 2 - 4\cos^2 y - \sin x$. The function ϕ can be expanded into an infinite series as

$$\phi = x \sum_{n=0}^{\infty} (-1)^n \frac{x^{2n}}{(2n+1)!}$$

$$+ yx + 10$$

$$+ (1 + \frac{1}{2} \sum_{k=1}^{\infty} (-1)^k \frac{(2y)^{2k}}{(2k)!})$$

or

$$\phi = 11 + x + yx - \frac{x^3}{3!} + \frac{x^5}{5!} - \frac{x^7}{7!} + \cdots$$

$$- y^2 + \frac{y^4}{3} - \frac{2y^6}{45} + \cdots$$

Table 6.6 lists the approximation functions for different evaluation point densities and basis function sets. The maximum relative error along the boundary is of the magnitude of order 10^{-4}. In all cases, $\varepsilon = 0.5$.

Table 6.6. Best Approximation Results for

Example 6.3.5

number of evaluation points Γ, Ω	$\hat{\phi}$
8, 16	$11 + .999x + .049y + xy - .013x^2 - 1.29y^2 + 0x^2y + 0xy^2 - .14\,x^3 + .53y^3$
8, 25	$11 + .996x + .044y + xy - .011x^2 - 1.28y^2 + 0x^2y + 0xy^2 - .14x^3 + .53y^3$
16, 16	$11 + x + .047y + xy - .016x^2 - 1.39y^2 + .0009x^2y - .000064xy^2 - .14x^3 + .53y^3$
16, 25	$11 + .998x + .041y + xy - .013x^2 - 1.28y^2 - .14x^3 + .53y^3$
16, 25	$11 + .9992x + .0008654y + .9995xy + .000607x^2 - 1.07y^2 - .1837x^3 + .1806y^3 + 1.986x^4 + 1.729y^4$

6.4. Computer Program: Two–Dimensional Potential Problems Using Real Variable Basis Functions

6.4.1 Introduction

The generalized Fourier series analysis program (GFSA1.FOR) consists of a main program and the subroutine FBASIS1.FOR. Figure 6.11 shows the simple flow chart of the GFSA1 program. The subroutine FBASIS1 consists of the selected basis functions and their Laplacian relationship. The last section of subroutine FBASIS1 consists of the approximation function calculations for boundary values and the interior Laplacian relationship. The user can enter up to 20 basis functions. FORTRAN listings of the programs are included in the following section. The basis functions shown in subroutine FBASIS1 are

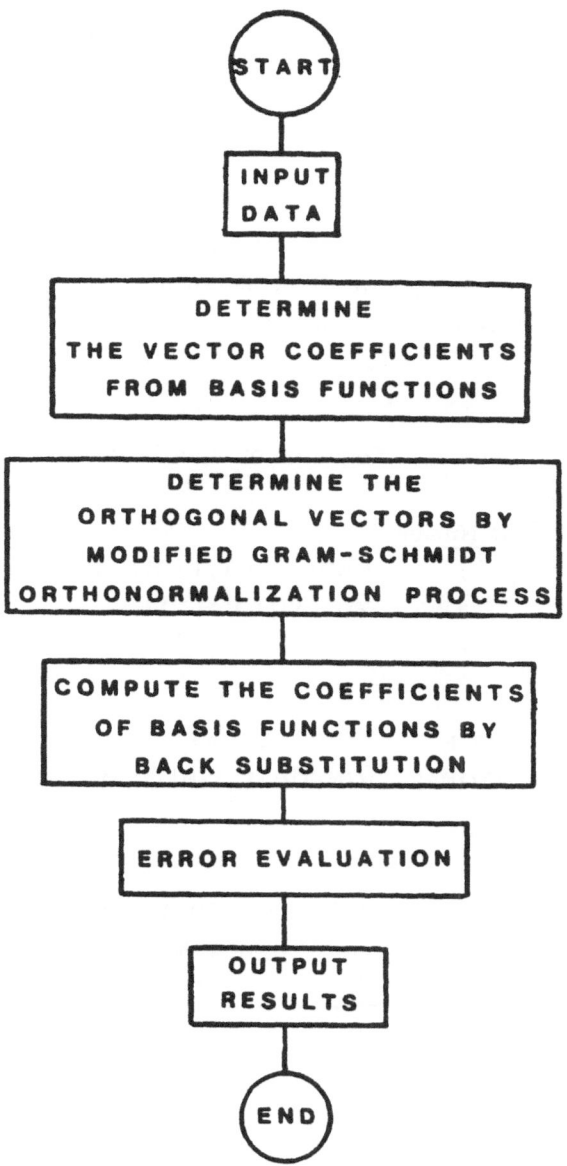

Fig. 6.11. Simple Flow Chart for Program GFSA1.

1, x, y, xy, x^2, y^2, x^2y, xy^2, x^3, y^3. The provided Laplacian relationships are 0, 0, 0, 0, $2T_{xx}$, $2T_{yy}$, $2yT_{xx}$, $2xT_{yy}$, $6xT_{xx}$, $6yT_{yy}$ according to $L(\phi) = T_{xx} \dfrac{\partial^2\phi}{\partial x^2} + T_{yy} \dfrac{\partial^2\phi}{\partial y^2}$.

6.4.2 Input Data Description

The input data are stored in data file—GFSA1.DAT. The input data should be entered sequentially as shown in Table 6.7. Table 6.8 shows the descriptions of input variables.

Table 6.7. Sequence of Input Data

Type of Data	Line Number	Variable
Control Data	1	NBON, NINT, EPSLO, KODE
	2	X(1), Y(1), VALUE(1)
Boundary	·	
	·	
Nodes	·	
	NBON+1	X(NBON), Y(NBON), VALUE(NBON)
	NBON+2	X(NBON+1), Y(NBON+1), VALUE(NBON+1)
Interior	·	
	·	
Nodes	·	
	NBON+1+NINT	X(NBON+NINT), Y(NBON+NINT), VALUE(NBON+NINT)

Table 6.8. Description of Input Variables

Variable	Description
NBON	Total number of nodes along domain boundary,
NINT	Total number of nodes in the domain,
EPSLO	Weighting factor for inner product (between 0 and 1)
KODE	Output options $\begin{cases} 0 - \text{Summary of results} \\ 1 - \text{detail of results} \end{cases}$
X(I)	Array stores x-coordinate for boundary and interior nodes
Y(I)	Array stores y-coordinate for boundary and interior nodes
VALUE(I)	Array stores boundary values and source term for interior nodal points

6.4.3. Computer Program Listing

```
C
C          MAIN PROGRAM
C
C          THIS IS A GENERALIZED FOURIER SERIES ANALYSIS PROGRAM
C          WHICH SOLVES THE POISSON EQUATION
C
           IMPLICIT DOUBLE PRECISION(A-H,O-Z)
           COMMON/BLK 1/ X(200),Y(200)
           COMMON/BLK 2/ VALUE(200)
           COMMON/BLK 3/ B(20),S(20)
           COMMON/BLK 4/ G(20,200)
           COMMON/BLK 5/ XK(20,20)
C
C          OPEN DATA FILES
C
C.......INPUT DATA FILE  = GFSA.DAT
C.......OUTPUT DATA FILE = GFSA.ANS
C
           NRD=1
           NWT=2
           OPEN(UNIT=NRD,FILE='GFSA.DAT',STATUS='OLD')
           OPEN(UNIT=NWT,FILE='GFSA.ANS',STATUS='NEW')
C
C          READ INPUT DATA
C
C.......DEFINITION OF VARIABLES
C
C          NBON : TOTAL NODE NUMBER ON THE BOUNDARY
C          NINT : TOTAL NODE NUMBER ON THE INTERIOR
C          EPSLO: WEIGHTING FACTOR FOR THE INNER PRODUCT
C          KODE : 0 - SUMMARY OF RESULTS
C                 1 - DETAIL OF RESULTS
C          X(I), Y(I) : X- AND Y- COORDINATES FOR NODE I
C          VALUE(I)   : BOUNDARY VALUE OR VALUE OF SOURCE TERM FOR NODE I
C
           READ(NRD,*) NBON,NINT,EPSLO,KODE
           DO 7 I=1,NBON
7          READ(NRD,*) X(I),Y(I),VALUE(I)
           WRITE(NWT,2)
           WRITE(NWT,17)EPSLO
C
C          OUTPUT FORMATS
C
2          FORMAT(//,10X,'*** INFORMATION OF BOUNDARY NODES ***',
      1    //,' NODE   X(I)        Y(I)          VALUE(I)')
3          FORMAT(2X,2I3,3X,D10.4)
4          FORMAT(/,10X,'*** APPROXIMATE INTERIOR NODAL VALUES AND ERRORS'
      1    ,' ***',//,6X,'NODE',9X,'EXACT',9X,'APPROXIMATION',5X,
      2    'RELATIVE ERROR',8X,'F(X,Y)',6X,'ESTIMATED F(X,Y)',3X,
      3    'RELATIVE ERROR')
5          FORMAT(3X,I5,3(8X,D10.4))
8          FORMAT(1X,I3,2X,F8.3,2X,F8.3,4X,F7.2)
17         FORMAT(/,'WEIGHTING FACTOR FOR INNER PRODUCT = ',F4.2/)
22         FORMAT(10(1X,D10.4))
```

```
23        FORMAT(/,10X,'*** NODE # ',I3,' ***')
24        FORMAT(/,10X,'*** NODAL POINT VECTOR EXPANSION, F(I) ***'/)
25        FORMAT(/,10X,'*** ORTHOGONAL VECTOR EXPANSION, G(I) ***'/)
26        FORMAT(/,10X,'*** ORTHOGONAL TEST (G(I),G(J)) ***',//,
     1    '   I  J  (G(I),G(J))')
28        FORMAT(/,10X,'EVALUATION COEFFICIENTS',
     1    ' (G(I),B(I))/(G(I),G(I)) ***'/)
29        FORMAT(/,10X,'*** BACK SUBSTITUTION COEFFICIENTS ***'/)
36        FORMAT(/,2X,'*** BESSEL INEQUALITY ***',/,2X,D10.4,
     1    ' >= ',D10.4,' AND THE DIFFERENCE IS ',D10.4,/,120('-')/)
38        FORMAT(2(1X,F10.4),1X,I2,3(1X,F10.4))
41        FORMAT(//,10X,'*** INFORMATION OF INTERIOR POINTS ***',
     1    //,' POINT  X(I)        Y(I)         VALUE(I)')
45        FORMAT(/,10X,'*** APPROXIMATE BOUNDARY VALUES AND ERRORS ***',
     1    //,5X,'POINT',9X,'EXACT',9X,'APPROXIMATION',5X,'RELATIVE',
     1    ' ERROR'/)
57        FORMAT(3X,I5,6(8X,D10.4))
81        FORMAT(1X,I3,2X,F8.3,2X,F8.3,4X,F7.2)
99        FORMAT(/,120('='))
C
C         INITIALIZE CONSTANTS
C
          NTOL=NBON
          SAREA=0.
          EX=0.
C
C         READ BOUNDARY INFORMATION FOR BOUNDARY NODES
C
          DO 300 I=1,NBON
          WRITE(NWT,8) I,X(I),Y(I),VALUE(I)
          SAREA=SAREA+EPSLO*VALUE(I)*VALUE(I)
300       CONTINUE
C
C         READ X-COORDINATE, Y-COORDINATE, AND THE SOURCE TERM
C         FOR THE INTERIOR POINTS
C
          WRITE(NWT,41)
          DO 305 I=1,NINT
          NTOL=NTOL+1
          READ (NRD,*) X(NTOL),Y(NTOL),VALUE(NTOL)
          SAREA=SAREA+VALUE(NTOL)*VALUE(NTOL)*(1.-EPSLO)
          WRITE(NWT,81)NTOL,X(NTOL),Y(NTOL),VALUE(NTOL)
305       CONTINUE
C
C         DETERMINE THE VECTOR COFFICIENTS FROM BASIS FUNCTIONS
C
          IF(KODE.EQ.1)WRITE(NWT,99)
          IF(KODE.EQ.1)WRITE(NWT,24)
          DO 315 I=1,NTOL
          XX=X(I)
          YY=Y(I)'
          CALL BASIS(NBAS,XX,YY,I,NBON,VBAR,FBAR,0)
315       CONTINUE
```

```
          IF(KODE.NE.1)GOTO 321
          DO 319 I=1,NBAS
          WRITE(NWT,22)(G(I,J),J=1,NTOL)
319       CONTINUE
C
C         USE THE MODIFIED GRAM-SCHMIDT ORTHONORMALIZATION PROCESS
C         TO DETERMINE SERIES OF ORTHOGONAL VECTORS
C
321       DO 320 I=1,NBAS
          SUM=0.
          DO 323 J=1,NTOL
          IF(J.LE.NBON)SUM=SUM+EPSLO*G(I,J)*G(I,J)
          IF(J.GT.NBON)SUM=SUM+(1.-EPSLO)*G(I,J)*G(I,J)
323       CONTINUE
          SUM=SQRT(SUM)
          S(I)=SUM
          IF(SUM.LT..0000001)S(I)=0.
          DO 325 J=1,NTOL
          IF(SUM.GT.0.)G(I,J)=G(I,J)/SUM
          IF(SUM.EQ.0.)G(I,J)=0.
325       CONTINUE
          IF(I.EQ.NBAS)GOTO 320
          IP1=I+1
          DO 310 KK=IP1,NBAS
          SUM1=0.
          DO 330 J=1,NTOL
          IF(J.GT.NBON)GOTO 335
          SUM1=SUM1+G(I,J)*G(KK,J)*EPSLO
          GO TO 330
335       SUM1=SUM1+(1.-EPSLO)*G(KK,J)*G(I,J)
330       CONTINUE
          XXK=-1.*SUM1
          IF(S(I).GT.0.)XK(KK,I)=XXK/S(I)
          IF(S(I).EQ.0.)XK(KK,I)=0.
          DO 340 J=1,NTOL
          G(KK,J)=G(KK,J)+XXK*G(I,J)
340       CONTINUE
310       CONTINUE
320       CONTINUE
          IF(KODE.NE.1.)GO TO 55
          WRITE(NWT,25)
          DO 47 I=1,NBAS
47        WRITE(NWT,22)(G(I,J),J=1,NTOL)
C.......CHECK ORTHOGONALITY OF VECTORS G(I)
          WRITE(NWT,26)
55        DO 50 I=1,NBAS
          IP1=I+1
          IF(I.EQ.NBAS)GO TO 80
          DO 70 K=IP1,NBAS
          SUM=0.
          DO 60 J=1,NTOL
          IF(J.LE.NBON)SUM=SUM+G(I,J)*G(K,J)*EPSLO
          IF(J.GT.NBON)SUM=SUM+G(I,J)*G(K,J)*(1.-EPSLO)
60        CONTINUE
```

```
        IF(KODE.EQ.1)WRITE(NWT,3)I,K,SUM
70      CONTINUE
50      CONTINUE
80      CONTINUE
C.......COMPUTE THE COEFFICIENTS OF B(I)=(VALUE(I),G(I))
        WRITE(NWT,28)
        SUM=0.
        DO 120 I=1,NBAS
        BK1=0.
        DO 130 J=1,NTOL
        IF(J.LE.NBON)BK1=BK1+VALUE(J)*G(I,J)*EPSLO
        IF(J.GT.NBON)BK1=BK1+VALUE(J)*G(I,J)*(1.-EPSLO)
130     CONTINUE
C.......COMPUTE THE NORM OF THE GENERALIZED FOURIER COEFFICIENTS
        IF(S(I).GT.0.)B(I)=BK1/S(I)
        IF(S(I).EQ.0.)B(I)=0.
        SUM=SUM+BK1*BK1
120     CONTINUE
        WRITE(NWT,22)(B(I),I=1,NBAS)
C.......COMPUTE BASIS FUNCTION COEFFICIENTS (BACK-SUBSTITUTION)
660     DO 200 I=NBAS,1,-1
        IF(I.EQ.NBAS)XK(NBAS,I)=B(NBAS)
        IF(I.NE.NBAS)XK(NBAS,I)=XK(NBAS,I)*B(NBAS)+B(I)
200     CONTINUE
        NTOT1=NBAS-1
        DO 210 I=NTOT1,1,-1
        DO 210 J=I,1,-1
        IF(I.EQ.J)GO TO 210
        IF(I.NE.J)XK(NBAS,J)=XK(NBAS,I)*XK(I,J)+XK(NBAS,J)
210     CONTINUE
        WRITE(NWT,29)
        WRITE(NWT,22)(XK(NBAS,I),I=1,NBAS)
C
C       BOUNDARY AND INTERIOR VALUES APPROXIMATION
C
C.......APPROXIMATE BOUNDARY POINT VALUES
        WRITE(NWT,99)
        WRITE(NWT,45)
        DO 470 I=1,NBON
        XX=X(I)
        YY=Y(I)
        CALL BASIS(NBAS,XX,YY,I,NBON,VBAR,FBAR,1)
        IF(VALUE(I).EQ.0.)XD=-9999.
        IF(VALUE(I).NE.0.)XD=(VALUE(I)-VBAR)/VALUE(I)
        WRITE(NWT,5)I,VALUE(I),VBAR,XD
470     CONTINUE
C.......APPROXIMATE INTERIOR POINT VALUES
        WRITE(NWT,99)
        WRITE(NWT,4)
        DO 280 I=NBON+1,NTOL
        CALL BASIS(NBAS,X(I),Y(I),I,NBON,VBAR,FBAR,1)
        EX=0.
```

```
        IF(EX.NE.O.)XD=(EX-VBAR)/EX
        IF(EX.EQ.O.)XD=-9999.
        IF(VALUE(I).EQ.O.)FD=-9999.
        IF(VALUE(I).NE.O.)FD=(VALUE(I)-FBAR)/VALUE(I)
        WRITE(NWT,57)I,EX,VBAR,XD,VALUE(I),FBAR,FD
280     CONTINUE
        WRITE(NWT,99)
C.......BESSEL'S INEQUALITY
        DIFF=SAREA-SUM
        IF(ABS(DIFF).LT.0.00001)DIFF=0.
        WRITE(NWT,36)SAREA,SUM,DIFF
        STOP
        END
```

```
C****************************************************************
          SUBROUTINE BASIS(NB,XX,YY,I,NNOD,VBAR,FBAR,KK)
C
C         THIS SUBROUTINE EVALUATES THE BASIS FUNCTION VALUES AND
C         THE LAPLACIAN OF THE BASIS FUNCTION VALUES
C
          IMPLICIT DOUBLE PRECISION (A-H,O-Z)
          COMMON/BLK 4/ G(20,200)
          COMMON/BLK 5/ XK(20,20)
          NB=10
          TXX=1.
          TYY=1.
          X2=XX*XX
          Y2=YY*YY
          XY=XX*YY
          IF(KK.EQ.1)GOTO 200
          IF(I.GT.NNOD)GOTO 100
C.......EVALUTE BASIS FUNCTIONS
          G(1,I)=1.
          G(2,I)=XX
          G(3,I)=YY
          G(4,I)=XY
          G(5,I)=X2
          G(6,I)=Y2
          G(7,I)=X2*YY
          G(8,I)=XX*Y2
          G(9,I)=XX*X2
          G(10,I)=YY*Y2
          RETURN
C.......EVALUTE LAPLACIAN OF BASIS FUNCTION
100       G(1,I)=0.
          G(2,I)=0.
          G(3,I)=0.
          G(4,I)=0.
          G(5,I)=2.*TXX
          G(6,I)=2.*TYY
          G(7,I)=2.*YY*TXX
          G(8,I)=2.*XX*TYY
          G(9,I)=6.*XX*TXX
          G(10,I)=6.*YY*TYY
          RETURN
C.......EVALUTE APPROXIMATION FUNCTION (VBAR)
C.......AND LAPLACIAN OF APPROXIMTION FUNCTION (FBAR)
200       VBAR=XK(NB,1)+XK(NB,2)*XX+XK(NB,3)*YY+XK(NB,4)*XY+XK(NB,5)*X2
     1    +XK(NB,6)*Y2+XK(NB,7)*X2*YY+XK(NB,8)*XX*Y2+XK(NB,9)*XX*X2
     1    +XK(NB,10)*YY*Y2
          FBAR=2.*XK(NB,5)*TXX+2.*XK(NB,7)*YY*TXX
     1    +6.*XK(NB,9)*XX*TXX
     1    +2.*XK(NB,6)*TYY+2.*XK(NB,8)*XX*TYY
     1    +6.*XK(NB,10)*YY*TYY
          RETURN
          END
```

Example 6.4.1

The potential function for ideal fluid flow around a 90° corner is studied in this example (see Example 6.2.1).

Input Data

```
16 16 .5 0
0.00 0.00 0.00
0.25 0.00 0.0625
0.50 0.00 0.25
0.75 0.00 0.5625
1.00 0.00 1.00
1.00 0.25 0.9375
1.00 0.50 0.75
1.00 0.75 0.4375
1.00 1.00 0.00
0.75 1.00 -.4375
0.50 1.00 -.75
0.25 1.00 -.9375
0.00 1.00 -1.0
0.00 0.75 -.5625
0.00 0.50 -.25
0.00 0.25 -.0625
0.20 0.20 0.00
0.20 0.40 0.00
0.20 0.60 0.00
0.20 0.80 0.00
0.40 0.20 0.00
0.40 0.40 0.00
0.40 0.60 0.00
0.40 0.80 0.00
0.60 0.20 0.00
0.60 0.40 0.00
0.60 0.60 0.00
0.60 0.80 0.00
0.80 0.20 0.00
0.80 0.40 0.00
0.80 0.60 0.00
0.80 0.80 0.00
```

Summary Results

*** INFORMATION OF BOUNDARY NODES ***

NODE   X(I)   Y(I)   VALUE(I)

WEIGHTING FACTOR FOR INNER PRODUCT = .50

1	.000	.000	.00
2	.250	.000	.06
3	.500	.000	.25
4	.750	.000	.56
5	1.000	.000	1.00
6	1.000	.250	.94
7	1.000	.500	.75
8	1.000	.750	.44
9	1.000	1.000	.00
10	.750	1.000	.-.44
11	.500	1.000	-.75
12	.250	1.000	-.94
13	.000	1.000	-1.00
14	.000	.750	-.56
15	.000	.500	-.25
16	.000	.250	-.06

*** INFORMATION OF INTERIOR POINTS ***

POINT   X(I)   Y(I)   VALUE(I)

17	.200	.200	.00
18	.200	.400	.00
19	.200	.600	.00
20	.200	.800	.00
21	.400	.200	.00
22	.400	.400	.00
23	.400	.600	.00
24	.400	.800	.00
25	.600	.200	.00
26	.600	.400	.00
27	.600	.600	.00
28	.600	.800	.00
29	.800	.200	.00
30	.800	.400	.00
31	.800	.600	.00
32	.800	.800	.00

EVALUATION COEFFICIENTS (G(I),B(I))/(G(I),G(I)) ***

.1472D-16 .1000D+01 -.1000D+01 .1550D-14 .4140D-02 -.1000D+01 -.1042D-15 -.1644D-15 .5267D-14 .2448D-14

*** BACK SUBSTITUTION COEFFICIENTS ***

.1855D-14 -.1881D-14 -.1031D-13 .2464D-13 .1000D+01 -.1000D+01 -.7373D-14 -.1581D-13 .5267D-14 .2448D-14

*** APPROXIMATE BOUNDARY VALUES AND ERRORS ***

POINT	EXACT	APPROXIMATION	RELATIVE ERROR
1	.0000D+00	.1855D-14	-.9999D+04
2	.6250D-01	.6250D-01	-.1754D-13
3	.2500D+00	.2500D+00	-.2220D-15
4	.5625D+00	.5625D+00	.1184D-14
5	.1000D+01	.1000D+01	.7772D-15
6	.9375D+00	.9375D+00	-.4737D-15
7	.7500D+00	.7500D+00	-.7401D-15
8	.4375D+00	.4375D+00	-.1269D-15
9	.0000D+00	-.1053D-14	-.9999D+04
10	-.4375D+00	-.4375D+00	.0000D+00
11	-.7500D+00	-.7500D+00	.1184D-14
12	-.9375D+00	-.9375D+00	.1184D-14
13	-.1000D+01	-.1000D+01	.1110D-15
14	-.5625D+00	-.5625D+00	-.2566D-14
15	-.2500D+00	-.2500D+00	-.5773D-14
16	-.6250D-01	-.6250D-01	-.4885D-14

*** APPROXIMATE INTERIOR NODAL VALUES AND ERRORS ***

NODE	EXACT	APPROXIMATION	RELATIVE ERROR	F(X,Y)	ESTIMATED F(X,Y)	RELATIVE ERROR
17	.0000D+00	.2832D-15	-.9999D+04	.0000D+00	.2083D-15	-.9999D+04
18	-.1200D+00	-.1200D+00	-.3123D-14	.0000D+00	.1962D-15	-.9999D+04
19	-.3200D+00	-.3200D+00	-.1735D-14	.0000D+00	.1843D-15	-.9999D+04
20	-.6000D+00	-.6000D+00	-.1850D-15	.0000D+00	.1723D-15	-.9999D+04
21	.1200D+00	.1200D+00	-.1503D-14	.0000D+00	.2065D-15	-.9999D+04
22	.0000D+00	-.5138D-16	-.9999D+04	.0000D+00	.1945D-15	-.9999D+04
23	-.2000D+00	-.2000D+00	-.2776D-15	.0000D+00	.1826D-15	-.9999D+04
24	-.4800D+00	-.4800D+00	.4626D-15	.0000D+00	.1705D-15	-.9999D+04
25	.3200D+00	.3200D+00	.0000D+00	.0000D+00	.2046D-15	-.9999D+04
26	.2000D+00	.2000D+00	-.2776D-15	.0000D+00	.1926D-15	-.9999D+04
27	.0000D+00	.1092D-15	-.9999D+04	.0000D+00	.1807D-15	-.9999D+04
28	-.2800D+00	-.2800D+00	.7930D-15	.0000D+00	.1687D-15	-.9999D+04
29	.6000D+00	.6000D+00	.1850D-15	.0000D+00	.2027D-15	-.9999D+04
30	.4800D+00	.4800D+00	-.4626D-15	.0000D+00	.1907D-15	-.9999D+04
31	.2800D+00	.2800D+00	-.7930D-15	.0000D+00	.1789D-15	-.9999D+04
32	.0000D+00	.2268D-16	-.9999D+04	.0000D+00	.1669D-15	-.9999D+04

*** BESSEL INEQUALITY ***
.3016D+01 >= .3016D+01 AND THE DIFFERENCE IS .0000D+00

In this example, since we know the exact solution of the problem, additional computer statements are included in the program to evaluate the relative error for the Laplacian relationship of the interior points.

Example 6.4.2

The soil-water flow through a homogeneous soil is studied in this example (see Example 6.2.6).

Input Data

```
17 14 .5 0
0. 0. 24.0
5. 0. 18.75
11. 0. 11.5
16. 0. 4.0
16. 4. 4.0
16. 8. 8.0
16. 12.62 12.62
14. 17.75 17.75
12. 19.86 19.86
9. 21.71 21.71
6. 23.02 23.02
3. 23.5 24.
0. 24. 24.
0. 20. 24.
0. 15. 24.
0. 10. 24.
0. 5. 24.
4. 4. 0.
4. 8. 0.
4. 12. 0.
4. 16. 10.
4. 20. 0.
8. 4. 0.
8. 8. 0.
8. 12. 0.
8. 16. 0.
8. 20. 0.
12. 4. 0.
12. 8. 0.
12. 12. 0.
12. 16. 0.
```

Summary Results

*** INFORMATION OF BOUNDARY NODES ***

NODE   X(I)   Y(I)   VALUE(I)

WEIGHTING FACTOR FOR INNER PRODUCT = .50

NODE	X(I)	Y(I)	VALUE(I)
1	.000	.000	24.00
2	5.000	.000	18.75
3	11.000	.000	11.56
4	16.000	.000	4.00
5	16.000	4.000	4.00
6	16.000	8.000	8.00
7	16.000	12.620	12.62
8	14.000	17.750	17.75
9	12.000	19.860	19.86
10	9.000	21.710	21.71
11	6.000	23.020	23.02
12	3.000	23.500	24.00
13	.000	24.000	24.00
14	.000	20.000	24.00
15	.000	15.000	24.00
16	.000	10.000	24.00
17	.000	5.000	24.00

*** INFORMATION OF INTERIOR POINTS ***

POINT	X(I)	Y(I)	VALUE(I)
18	4.000	4.000	.00
19	4.000	8.000	.00
20	4.000	12.000	.00
21	4.000	16.000	.00
22	4.000	20.000	.00
23	8.000	4.000	.00
24	8.000	8.000	.00
25	8.000	12.000	.00
26	8.000	16.000	.00
27	8.000	20.000	.00
28	12.000	4.000	.00
29	12.000	8.000	.00
30	12.000	12.000	.00
31	12.000	16.000	.00

EVALUATION COEFFICIENTS (G(I),B(I))/(G(I),G(I)) ***

.1819D+02 -.9360D+00 .2876D+00 .4784D-01 -.2360D-01 .2178D-02 .1155D-02 .1484D-02 .1990D-02 -.3129D-03

*** BACK SUBSTITUTION COEFFICIENTS ***

.2400D+02 -.6549D+00 -.9890D-01 -.1864D-01 -.6852D-01 .1180D-01 .2409D-02 .1676D-02 .1878D-02 -.3129D-03

*** APPROXIMATE BOUNDARY VALUES AND ERRORS ***

POINT	EXACT	APPROXIMATION	RELATIVE ERROR
1	.2400D+02	.2400D+02	.6667D-04
2	.1875D+02	.1925D+02	-.2644D-01
3	.1156D+02	.1100D+02	.4813D-01
4	.4000D+01	.3673D+01	.8185D-01
5	.4000D+01	.5149D+01	-.2874D+00
6	.8000D+01	.7742D+01	.3222D-01
7	.1262D+02	.1197D+02	.5163D-01
8	.1775D+02	.1791D+02	-.9201D-02
9	.1986D+02	.2014D+02	-.1424D-01
10	.2171D+02	.2185D+02	-.6226D-02
11	.2302D+02	.2292D+02	.4224D-02
12	.2400D+02	.2358D+02	.1770D-01
13	.2400D+02	.2410D+02	-.4130D-02
14	.2400D+02	.2424D+02	-.9970D-02
15	.2400D+02	.2412D+02	-.4792D-02
16	.2400D+02	.2388D+02	.5125D-02
17	.2400D+02	.2376D+02	.1000D-01

*** APPROXIMATE INTERIOR NODAL VALUES AND ERRORS ***

NODE	EXACT	APPROXIMATION	RELATIVE ERROR	F(X,Y)	ESTIMATED F(X,Y)	RELATIVE ERROR
18	.0000D+00	.2014D+02	-.9999D+04	.0000D+00	-.4318D-01	-.9999D+04
19	.0000D+00	.2035D+02	-.9999D+04	.0000D+00	-.3141D-01	-.9999D+04
20	.0000D+00	.2091D+02	-.9999D+04	.0000D+00	-.1964D-01	-.9999D+04
21	.0000D+00	.2170D+02	-.9999D+04	.0000D+00	-.7876D-02	-.9999D+04
22	.0000D+00	.2261D+02	-.9999D+04	.0000D+00	.3890D-02	-.9999D+04
23	.0000D+00	.1534D+02	-.9999D+04	.0000D+00	.1532D-01	-.9999D+04
24	.0000D+00	.1604D+02	-.9999D+04	.0000D+00	.2708D-01	-.9999D+04
25	.0000D+00	.1730D+02	-.9999D+04	.0000D+00	.3885D-01	-.9999D+04
26	.0000D+00	.1901D+02	-.9999D+04	.0000D+00	.5062D-01	-.9999D+04
27	.0000D+00	.2104D+02	-.9999D+04	.0000D+00	.6238D-01	-.9999D+04
28	.0000D+00	.1011D+02	-.9999D+04	.0000D+00	.7381D-01	-.9999D+04
29	.0000D+00	.1160D+02	-.9999D+04	.0000D+00	.8558D-01	-.9999D+04
30	.0000D+00	.1387D+02	-.9999D+04	.0000D+00	.9734D-01	-.9999D+04
31	.0000D+00	.1680D+02	-.9999D+04	.0000D+00	.1091D+00	-.9999D+04

*** BESSEL INEQUALITY ***
.3242D+04 >= .3240D+04 AND THE DIFFERENCE IS .1494D+01

The relative error is defined as:

$$\text{Relative Error} = \begin{cases} \dfrac{|\text{ Exact} - \text{Approximation }|}{\text{Exact}} & ; \text{ Exact} \neq 0 \\[2ex] -999 & ; \text{ otherwise} \end{cases}$$

166

6.5. Application of Computer Program

The problem considered is the second order ordinary differential equation

$$y" + y = 0, \ y(0) = 0, \ y(\pi/2) = 1$$

Generally, the problem statement will be given in the form of a linear operator which describes the points within a boundary, and in the form of point values on the boundary. From the given set of conditions a vector, Φ, is constructed. The elements of this vector represent appropriate problem solving values at evaluation points which will be used in the approximation. A set of basis vectors, $\{F_i\}$, will be used in the least squares approximation of Φ. For example, to solve the example problem using 10 evaluation points on the interval $(0, \pi/2)$, then

$$\Phi = (0, 1, 0, 0,...,0)$$

where the first two entries, 0 and 1, are the values of y at the end points, 0 and $\pi/2$, and the remaining zeros represent the values of (y" + y) applied to the 10 evaluation points in the interior.

The interior evaluation points are chosen depending on where the most accuracy is desired. The more evaluation points that are chosen in a certain area, the more accurate will the approximation be in that area. The weighting factor for the boundary points is also determined by accuracy considerations. The inner product weighting factor, ε, is a number between 0 and 1 which scales the boundary point values. The factor $(1-\varepsilon)$ then scales the interior points. If the approximation is to be more accurate at the boundary points then ε is set close to 1, and if the interior points are more critical then ε is set near 0.

For the example problem y" + y = 0, consider the trial functions $\{1, x, \sqrt{x}, \sin x, \cos x, x\sin x, x\cos x\}$. The basis vectors $\{F_i\}$ are constructed as follows:

$$F_1 \to 1: [1,1,1,1,...,1]$$
$$F_2 \to x: [0, \pi/2,x,x,...,x]$$
$$F_3 \to \sqrt{x}: [0,\sqrt{\pi}/2,2+\sqrt{x},...,2+\sqrt{x}\ ]$$
$$F_4 \to \sin x: [0,1,0,0,...,0]$$
$$F_5 \to \cos x: [1,0,0,0,...,0]$$
$$F_6 \to x\sin x: [0, \pi/2,2\cos x,...,2\cos x]$$
$$F_7 \to x\cos x: [0,0,-2\sin x,...,-2\sin x]$$

The first element in each vector is the function evaluated at the first boundary point (x=0), the second element in each vector is the function evaluated at the second boundary point (x=π/2), and the remaining elements are the results of the operator Lf_i at the interior evaluation points.

6.5.1 A Fourth Order Differential Equation

Consider the boundary value problem:

$$y'''' + 2y'' + y = 0; \quad y(0) = 1, \quad y'(0) = 0, \quad y(\tfrac{\pi}{2}) = 0, \quad y'(\tfrac{\pi}{2}) = 1,$$

with solution $y(x) = 2.14 \sin x + \cos x - 2.14\, x\cos x - 1.363\, x\sin x$, $x \in [0, \pi/2]$. With a complete basis, the Best Approximation computer program provides the above exact solution. The following figures demonstrate the effect of an incomplete basis. In both cases, xsinx has been deleted from the set of basis functions. Figures 6.12 and 6.13 demonstrate the effect of higher order monomials being added to the set of basis functions, $\{f_i\}$. With the addition of each higher order monomial, the approximation becomes more accurate. Figures 6.14 and 6.15 demonstrate the effect of a change in inner product weighting, ε, i.e., focusing upon satisfying the problem's boundary conditions or upon satisfying the operator. As expected, a small ε satisfies the operator space well but gives a poor approximate solution, while a larger ε gives a good boundary approximation while performing poorly in the operator space.

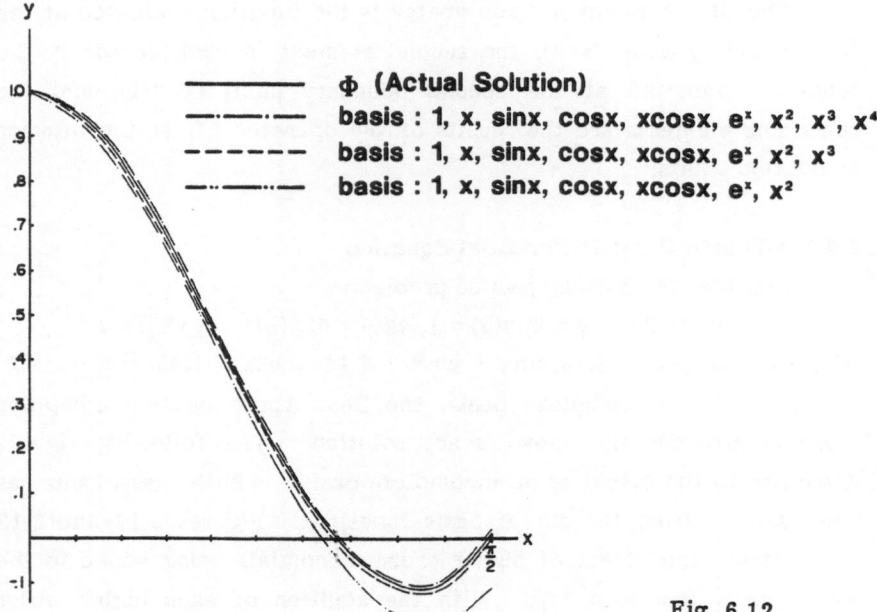

Fig. 6.12.

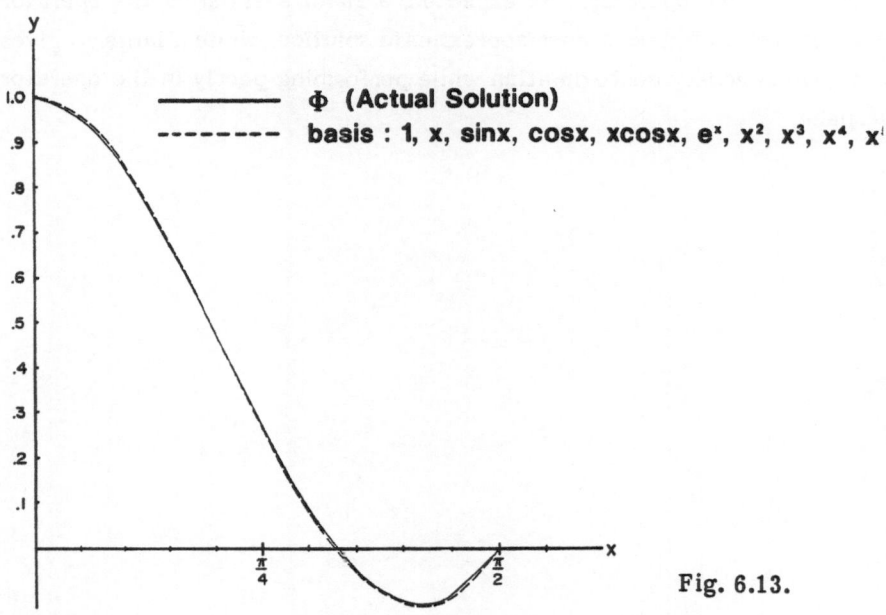

Fig. 6.13.

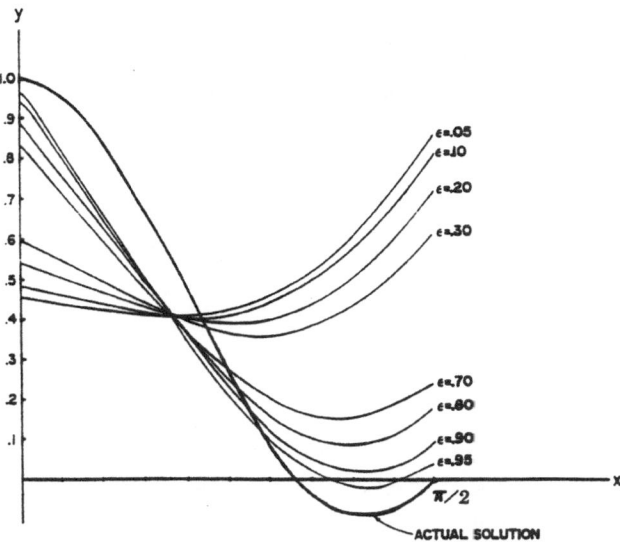

Fig. 6.14.

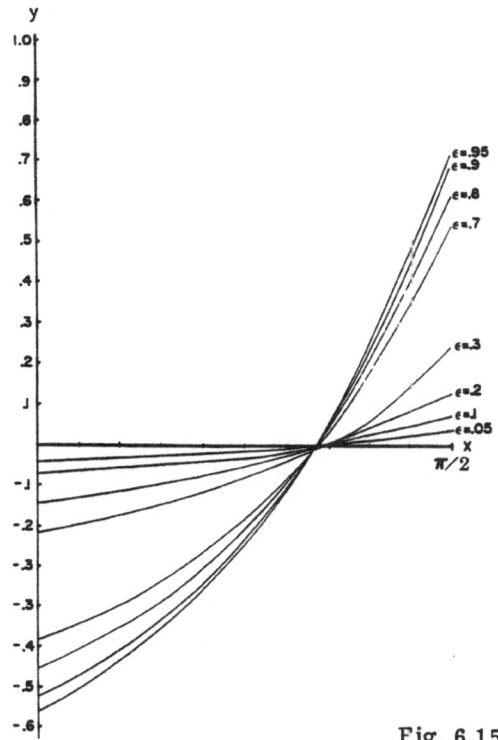

Fig. 6.15.

CHAPTER 7
SOLVING POTENTIAL PROBLEMS USING THE
BEST APPROXIMATION METHOD

7.0. Introduction

From Eq. (5.2.2), a linear operation equation such as involved in potential problems may be solved using the Best Approximation Method by use of the inner product

$$(u,v) = \int_\Gamma uvd\Gamma + \int_\Omega Lu \, Lv \, d\Omega \qquad (7.1.1)$$

where the integration over Γ includes both the spatial and temporal boundary conditions (i.e., initial conditions in a diffusion problem).

By a clever choice of basis functions, the inner product can be simplified. For example, choosing basis functions which satisfy the operator relationship ($Lu = Lv = 0$ in (7.1.1)) the inner product simplifies to

$$(u,v) = \int_\Gamma uvd\,\Gamma \qquad (7.1.2)$$

In this chapter, the two-dimensional Laplace equation (or Poisson equation) is examined with respect to using the Best Approximation Method where the set of basis functions are analytic functions. That is, an approximation function is developed which is the sum of complex variable analytic functions. Because only the boundary component of (7.1.1) is used in (7.1.2), the resulting approximator is a variant of the Complex Variable Boundary Element Method (Hromadka, 1984). Sections 7.1 through 7.6 focus upon the use of the specialized analytic functions for the two-dimensional basis functions.

The later sections of this chapter focuses upon some other families of basis functions, and their application to solving potential problems. Included in the topics considered are three-dimensional applications and temporal effects in flow regimes.

7.1. The Complex Variable Boundary Element Method

7.1.1. Objectives

The objective in using the Complex Variable Boundary Element Method (or CVBEM) is to approximate analytic complex functions. More specifically, if ω is a two-dimensional complex function which is analytic over a simply connected domain Ω with boundary values $\omega(\zeta)$ for $\zeta \in \Gamma$ (Γ is a simple closed contour), then the real (ϕ) and imaginary (ψ) parts of $\omega = \phi + i\psi$ both satisfy the Laplace equation over Ω. Thus, two-dimensional potential problems can be solved numerically by the CVBEM.

The development of the CVBEM for engineering applications is detailed in Hromadka (1984). Generally speaking, the CVBEM is a boundary integral technique. A literature review of this class of numerical methods can be found in other books such as Lapidus and Pinder (1982).

However in this chapter, the CVBEM departs from the other boundary integral methods by using a best approximation in satisfying boundary conditions. Instead of developing a square mxm matrix system for m boundary nodes by collocating the boundary integral equation at nodal point boundary values, the CVBEM is now expanded as a generalized Fourier series—eliminating the matrix solution entirely. Boundary conditions are approximated in a "mean-square" error sense in that a new vector space norm is defined which is analogous to the ℓ_2 norm, and then minimized by the selection of complex coefficients to be associated to each nodal point located with the problem boundary, Γ.

7.1.2. DEFINITION 7.1.1. (Working Space, W_Ω)

Let Ω be a simply connected convex domain with a simple closed piecewise linear boundary Γ and with its centroid located at $0 + 0i$. Then $\omega \in W_\Omega$ has the properties

(i) $\omega(z)$ is analytic over Ω

(ii) $\displaystyle \lim_{\delta \to 1} \int_\Gamma |\omega(\delta\zeta)|^2 \, d\Gamma \leq M < \infty$

7.1.3. DEFINITION 7.1.2. (the Function $||\omega||$)

A key element in the CVBEM development of this chapter is the definition of a norm and inner-product. In the following sections, insight into the new norm function is presented by an analogy to the well known $\ell_2(\Gamma)$ norm and inner-product.

The symbol $||\omega||$ for $\omega \in W_\Omega$ is notation for

$$||\omega|| = \left[\int_{\Gamma_\phi} (\text{Re}\omega)^2 d\mu + \int_{\Gamma_\psi} (\text{Im}\omega)^2 d\mu \right]^{\frac{1}{2}}$$

The symbol $||\omega||_p$ for $\omega \in W_\Omega$ is notation for

$$||\omega||_p = \left[\int_\Gamma |\omega(\zeta)|^p d\mu \right]^{1/p}, \ p \geq 1$$

Of importance is the case of $p = 2$:

$$||\omega||_2 = \left[\int_\Gamma |\omega(\zeta)|^2 d\mu \right]^{\frac{1}{2}} = \left[\int_\Gamma \left([\text{Re}\omega]^2 + [\text{Im}\omega]^2 \right) d\mu \right]^{\frac{1}{2}}$$

7.1.4. Almost Everywhere (ae) Equality

Because sets of Lebesque measure zero have no effect on integration, almost-everywhere (ae) equality on Γ indicates the same class of element. Thus for $\omega \in W_\Omega$, $[\omega] = \{\omega \in W_\Omega : \omega(\zeta)$ are equal ae for $\zeta \in \Gamma\}$. For example, $[0] = \{\omega \in W_\Omega : \omega(\zeta) = 0$ ae, $\zeta \in \Gamma\}$. When understood, the notation "[ ]" will be dropped.

7.1.5. Theorem (relationship of $||\omega||$ to $||\omega||_2$)

Let $\omega \in W_\Omega$. Then $||\omega||_2^2 = ||\omega||^2 + ||i\omega||^2$

Proof

Let $\omega = \phi + i\psi$.

Then $\|\omega\|_2^2 = \int_\Gamma |\omega(\varsigma)|^2 d\mu$

$$= \int_\Gamma (\phi^2 + \psi^2) d\mu$$

$$= \int_{\Gamma_\phi} \phi^2 d\mu + \int_{\Gamma_\psi} \phi^2 d\mu + \int_{\Gamma_\phi} \psi^2 d\mu + \int_{\Gamma_\psi} \psi^2 d\mu$$

$$= \|\omega\|^2 + \int_{\Gamma_\phi} (-\psi)^2 d\mu + \int_{\Gamma_\psi} \phi^2 d\mu$$

$$= \|\omega\|^2 + \|i\omega\|^2$$

7.1.6. Theorem

Let $\omega \in W_\Omega$. Then $\|\omega\|_2 = 0 \Rightarrow \|\omega\| = 0$.

Proof

$\|\omega\|_2 = 0$ implies $\|\omega\|_2^2 = \|\omega\|^2 + \|i\omega\|^2 = 0$.

7.1.7. Theorem

Let $\omega \in W_\Omega$. Then $\|\omega\| \leq \|\omega\|_2$.

Proof

Let $\omega = \phi + i\psi$. Then

$$||\omega||_2^2 = \int_\Gamma |\omega(\zeta)|^2 d\mu = \int_\Gamma \phi^2 d\mu + \int_\Gamma \psi^2 d\mu$$

$$= ||\omega||^2 + ||i\omega||^2$$

Because $||i\omega||^2 \geq 0$, then $||\omega||^2 \leq ||\omega||_2^2$.

7.2. Mathematical Development

7.2.1. Discussion: (A Note on Hardy Spaces)

The H^p spaces (or Hardy spaces) are well documented in the literature (e.g. Duren, 1970). Of special interest are the $E^p(\Omega)$ spaces of complex valued functions. If $\omega \in E^2(\Omega)$, then ω satisfies the conditions of Definition 1.6.1 for W_Ω, where $||\omega(\delta\zeta)||_2$ is bounded as $\delta \to 1$. Finally, if $\omega \in E^2(\Omega)$ then the Cauchy integral representation of $\omega(z)$ for $z\epsilon\Omega$ applies. It is seen that $W_\Omega \subset E^2(\Omega)$.

7.2.2. Theorem (Boundary Integral Representation)

Let $\omega \in W_\Omega$ and $z \in \Omega$. Then

$$\omega(z) = \frac{1}{2\tau i} \int_{\Gamma'} \frac{\omega(\zeta)d\zeta}{\zeta - z}$$

Proof

For $\omega \in W_\Omega$, then $\omega \in E^2(\Omega)$ and the result follows immediately.

7.2.3. Almost Everywhere (ae) Equivalence

For $\omega \in W_\Omega$, functions $x \in W_\Omega$ which are equal to ω ae on Γ represent an equivalence class of functions which may be noted as $[\omega]$.

Thus for x and y in W_Ω, x = y implies x = y over Ω and $\int_\Gamma |x - y| d\mu = 0$.

For simplicity, $\omega \in W_\Omega$ is understood to indicate $[\omega]$. This follows directly from the boundary integral representation of $\omega(z)$ for $z \in \Omega$, and the fact that integrals over sets of measure zero have no effect on the integral value.

7.2.4. Theorem (Uniqueness of Zero Element in W_Ω)

Let $\omega \in W_\Omega$ and $\phi = 0$ ae on Γ_ϕ and $\psi = 0$ ae on Γ_ψ. Then $\omega = [0] \in W_\Omega$.

Proof

Let $z \in \Omega$. Then

$$\omega(z) = \frac{1}{2\pi i} \int_\Gamma \frac{\omega(\zeta)d\zeta}{\zeta - z} = \frac{1}{2\pi i} \int_{\Gamma_\psi} \frac{\phi(\zeta)d\zeta}{\zeta - z} + \frac{1}{2\pi i} \int_{\Gamma_\phi} \frac{i\psi(\zeta)d\zeta}{\zeta - z}$$

due to $\phi(\zeta) = 0$ ae on Γ_ϕ and $\psi(\zeta) = 0$ ae on Γ_ψ. Therefore an equivalent function $\omega^* = \phi^* + i\psi^*$ can be formed where $\omega^* \in W_\Omega$ and

$$\phi^*(z) = \begin{cases} 0, & z \in \Gamma_\phi \\ \\ \phi(z), & z \in \Gamma_\psi \end{cases}$$

and $\omega(z) = \omega^*(z)$ for all $z \in \Omega$.

By the Riemann Mapping Theorem and Caratheodory's Extension of the Riemann Mapping Theorem, any two bounded simply connected domains can be conformally mapped onto each other with a one-to-one correspondence (from the continuous extension of the conformal mapping) of the boundary points. In a recent textbook, Mathews (1982) shows that for the case of ϕ^* being constant on Γ_ϕ and $\frac{\partial \phi^*}{\partial n} = 0$ on Γ_ψ that ω^* is a constant complex number over Ω. By continuity, $\phi^* = 0$ over Ω and, by a similar argument, $\psi^* = 0$ over Ω. Thus $\omega^* = 0$ over Ω and, consequently, $\omega = 0$ over Ω. Hence, $\omega = [0]$.

7.2.5. Theorem (W_Ω is a Vector Space)

W_Ω is a linear vector space over the field of real numbers.

Proof

This follows directly from the character of analytic functions. The zero element has already been noted by $[0]$ in Theorem 7.2.4.

7.2.6. Theorem (Definition of the Inner-Product)

Let $x, y \in W_\Omega$. Define a real-valued function (x,y) by

$$(x,y) = \int_{\Gamma_\Omega} \text{Re}(x)\,\text{Re}(y)\,d\mu + \int_{\Gamma_\psi} \text{Im}(x)\,\text{Im}(y)\,d\mu$$

Then $(\ ,\ )$ is an inner-product over W_Ω.

Proof

It is obvious that $(x,y) = (y,x)$; $(kx,y) = k(x,y)$ for k real; $(x+y,z) = (x,z) + (y,z)$; and $(x,x) \geq 0$. By theorem 7.2.4, $(x,x) = 0$ implies $\text{Re}x = 0$ ae on Γ_ϕ and $\text{Im}x = 0$ ae on Γ_ψ and $x = [0] \in W_\Omega$.

Three theorems follow immediately from the above.

7.2.7. Theorem (W_Ω is an Inner-Product Space)

For the defined inner-product, W_Ω is an inner-product space over the field of real numbers.

7.2.8. Theorem ($\|\omega\|$ is a Norm on W_Ω)

A norm is defined by $\|x\| = (x,x)^{1/2}$ for $x \in W_\Omega$.

7.2.9. Theorem

Let $x \in W_\Omega$ and $\|x\| = 0$. Then $x = [0]$.

7.3. **The CVBEM and** W_Ω

7.3.1. DEFINITION 7.3.1. (Angle Points)

Let the number of angle points of Γ be noted as Λ. By a nodal partition P_n of Γ, $m = \Lambda(n-1)$ nodes $\{zj\}$ are defined on Γ such that a node is located at each angle-point of Γ and the remaining nodes are distributed on Γ. Nodes are numbered sequentially in a counterclockwise direction along Γ. The scale of P_n is indicated by ℓ where $\ell = \max |z_{j+1} - z_j|$. Thus n nodes are equally spaced along each line segment.

7.3.2. DEFINITION 7.3.2. (Boundary Element)
A boundary element Γ_j is the line segment joining nodes z_j and z_{j+1}.

7.3.3. Theorem
Let P_n be defined on Γ. Then

$$\Gamma = \bigcup_{j=1}^{m} \Gamma_j$$

where $m = \Lambda(n-1)$

Proof
Follows from Γ being piecewise linear, and the construction of P_n.

7.3.4. DEFINITION 7.3.3. (Linear Basis Function)
A linear basis function $N_j(\zeta)$ is defined for $\zeta \in \Gamma$ by

$$N_j(\zeta) = \begin{cases} (\zeta - z_{j-1})/(z_j - z_{j-1}) & , \ \zeta \in \Gamma_{j-1} \\ (z_{j+1} - \zeta)/(z_{j+1} - z_j) & , \ \zeta \in \Gamma_j \\ 0 & , \ \zeta \notin \Gamma_{j-1} \cup \Gamma_j \end{cases}$$

The $N_j(\zeta)$ are real-valued and bounded as is verified by the following theorem.

7.3.5. Theorem
Let $N_j(\zeta)$ be defined as in definition 7.3.3 for node $z_j \in \Gamma$. Then $0 \leq N_j(\zeta) \leq 1$.

7.3.6. DEFINITION 7.3.4. (Global Trial Function)
Let P_n be defined on Γ with $m \geq \Lambda$ and scale ℓ. At each node z_j, define nodal values $\bar{\omega}_j = \bar{\phi}_j + i\bar{\psi}_j$ where $\bar{\phi}_j$ and $\bar{\psi}_j$ are real numbers. A global trial function $G_m(\zeta)$ is defined on Γ by

$$G_m(\zeta) = \sum_{j=1}^{m} N_j(\zeta)\bar{\omega}_j$$

7.3.7. Theorem

$G_m(\zeta)$ is continuous on Γ.

7.3.8. Discussion

As a result of $\omega(\zeta) \in L_2(\Gamma)$, then $\omega(\zeta)$ is measurable on Γ and for every $\varepsilon > 0$ there exists a continuous complex-valued function $g(\zeta)$ such that $||\omega(\zeta) - g(\zeta)||_1 < \varepsilon/2$. Choosing $G_m(\zeta)$ to approximate $g(\zeta)$ by $||G_m(\zeta) - g(\zeta)||_1 < \varepsilon/2$,
then $||\omega(\zeta) - G_m(\zeta)||_1 \leq ||\omega(\zeta) - g(\zeta)||_1 + ||g(\zeta) - G_m(\zeta)||_1 < \varepsilon$.

7.3.9. Theorem

Let $\omega \in W_\Omega$. For $\varepsilon > 0$ there exists a $G_m(\zeta)$ such that $||\omega(\zeta) - G_m(\zeta)||_1 < \varepsilon$.

Proof

Follows from the discussion in 7.3.8.

7.3.10. Discussion

The CVBEM approximation function $\hat{\omega}_m(z)$ is developed from the singular integral for a partition P_n of Γ by

$$\hat{\omega}_m(z) = \frac{1}{2\pi i} \int_\Gamma \frac{G_m(\zeta)d\zeta}{\zeta - z}, \quad z \in \Omega \qquad (7.3.1)$$

where $G_m(\zeta) = \sum_{j=1}^{m} N_j(\zeta)\bar{\omega}_j$ is the global trial function chosen to achieve $||\omega(\zeta) - G_m(\zeta)||_1 < \varepsilon$ for $\omega \in W_\Omega$ and $\varepsilon > 0$. Expanding G_m in the integrand gives

$$\hat{\omega}_m(z) = \sum_{j=1}^{m} \frac{1}{2\pi i} \int_\Gamma \frac{N_j(\zeta)\bar{\omega}_j d\zeta}{\zeta - z}, \quad z \in \Omega \qquad (7.3.2)$$

Appendix A shows that $\hat{\omega}_m(z)$ can be written as

$$\hat{\omega}_m(z) = R_1(z) + \sum_{j=1}^{m} c_j(z - z_j) \operatorname{Ln}(z - z_j), \quad z \in \Omega \qquad (7.3.3)$$

where $R_1(z)$ is a first degree complex polynomial resulting from the 2π-circuit along Γ about point z; the complex logarithm is with respect to point z (the branch cut is a ray originating from point $z \in \Omega$); and the c_j are complex constants $c_j = a_j + ib_j$ where the a_j and b_j are real numbers. The problem now can be restated as how to choose the best values for the c_j (and the $R_1(z)$ constants) such as to minimize a defined norm. Because ϕ is known only on Γ_ϕ and ψ is known only on Γ_ψ ($\omega = \phi + i\psi$), $||\omega - \hat{\omega}_m||_2$ is undefined or unknown. Therefore, the constants will be chosen to minimize the newly defined norm $||\omega - \hat{\omega}_m||$ where the goal is $||\omega - \hat{\omega}_m|| \to 0 \Rightarrow \omega_m(z) \to \omega(z)$ for all $z \in \Omega$.

For development purposes, the $\mathrm{Ln}(z - z_j)$ functions are replaced by $\mathrm{Ln}_j(z - z_j)$ functions where logarithm branch cuts are rays from each z_j which lie exterior of $\bar{\Omega} - \{z_j\}$ (see Appendix A).

Letting $R_1(z) = c_{m+1} + c_{m+2}z$, the CVBEM approximation used is now defined as

$$\hat{\omega}_m(z) = \sum_{j=1}^{m+2} c_j T_j \tag{7.3.4}$$

where

$$T_j = \begin{cases} (z - z_j) \, \mathrm{Ln}_j \, (z - z_j) \, , & j = 1, 2, \cdots, m \\ 1 + 0i \, , & j = m+1 \\ z \, , & j = m+2 \end{cases}$$

By the use of the $\mathrm{Ln}_j(z - z_j)$ functions, $\hat{\omega}_m(z)$ is analytic over $\bar{\Omega}$ except at the nodal points, and $\hat{\omega}_m(z)$ is continuous over $\bar{\Omega}$. In fact, $\hat{\omega}_m(z)$ is analytic everywhere except along the branch cuts.

If $c_j = a_j + ib_j$ is substituted into (7.3.4), the CVBEM approximation can be written with respect to real number coefficients γ_j as

$$\hat{\omega}_m(z) = \sum_{j=1}^{2(m+2)} \gamma_j f_j \tag{7.3.5}$$

where the f_j functions are given by

$$
\left.
\begin{aligned}
f_1 &= (z - z_1) \, Ln_1 \, (z - z_1) \\
f_2 &= if_1 \\
&\;\vdots \\
f_{2m-1} &= (z - z_m) \, Ln_m \, (z - z_m) \\
f_{2m} &= if_{2m-1} \\
f_{2m+1} &= 1 \\
f_{2m+2} &= i \\
f_{2m+3} &= z \\
f_{2(m+2)} &= iz
\end{aligned}
\right\}
\qquad (7.3.6)
$$

7.3.11. Theorem (Linear Independence of Nodal Expansion Functions)

The set of functions $\{(z - z_j) \, Ln_j \, (z - z_j), \ j = 1,2,\cdots,m+1\}$ are linearly independent.

Proof

Suppose the first m functions are linearly independent, but the $(m+1)$ - function is linearly dependent on the other m functions. Then for complex constants c_j,

$$
c_{m+1} (z - z_{m+1}) \, Ln_{m+1}(z - z_{m+1}) = \sum_{j=1}^{m} c_j(z - z_j) \, Ln_j(z - z_j)
$$

Taking the second derivative with respect to z gives

$$
\frac{c_{m+1}}{(z - z_{m+1})} = \sum_{j=1}^{m} \frac{c_j}{(z - z_j)} , \quad \text{for } z \neq z_k, \ k = 1,2,\cdots,m+1
$$

Rearranging terms, the above implies that

$$c_{m+1} \prod_{k=1}^{m} (z - z_k) = (z - z_{m+1}) \sum_{j=1}^{m} c_j \prod_{\substack{k=1 \\ k \neq j}}^{m} (z - z_k)$$

which is valid only if $c_k = 0$ for each $k = 1, 2, \cdots, m+1$.

7.3.12. Discussion

From the previous theorem, the set of functions $\{T_j\}$ of (7.3.4) are also linearly independent and, more importantly in this development, the $\{f_j\}$ are linearly independent with respect to the real number field. Thus for a given number m of nodes on Γ, the functions $\{f_j; j = 1, 2, \cdots, m\}$ forms a basis for the vector space spanned by the $\{f_j\}$, noted by $\hat{W}_\Omega^m$. In this notation, m indicates the number of nodes defined on Γ (always, m $\geq \Lambda$), and the hat indicates the CVBEM approximation function vector space.

The CVBEM objective is to choose a $\hat{\omega}_m \in \hat{W}_\Omega^m$ which minimizes $\|\omega - \hat{\omega}_m\|$ where $\omega \in W_\Omega$ and the nodes $\{z_j\}$ are fixed on Γ.

7.3.13. Theorem

Let $\omega \in W_\Omega$ and $z \in \Omega$. For every $\varepsilon > 0$ there exists a CVBEM approximation $\hat{\omega}_m$ such that $|\omega(z) - \hat{\omega}_m(z)| < \varepsilon$.

Proof

Let $d = \min\{|\zeta - z|, \zeta \in \Gamma\}$. Then for a global trial function $G_m(\zeta)$ defined on Γ

$$|\omega(z) - \hat{\omega}_m(z)| = \left| \frac{1}{2\pi i} \int_{\Gamma} \frac{[\omega(\zeta) - G_m(\zeta)]d\zeta}{\zeta - z} \right|$$

$$\leq \frac{1}{2\pi d} \|\omega - G_m\|_1 \leq \frac{\sqrt{L}}{2\pi d} \|\omega - G_m\|_2$$

Choosing G_m (see secton 7.3.10) such that $||\omega - G_m||_2 < 2\pi d \; \varepsilon/\sqrt{L}$ (or $||\omega - G_m||_1 < 2\pi d \; \varepsilon$) guarantees the desired result.

More insight as to the power of the CVBEM is provided by an analogy to convergence in measure.

7.3.14. Theorem

Let $\varepsilon > 0$. Then there exists a $0 < \delta < 1$ such that the

$$\int\!\!\!\int\limits_{\Omega - \Omega_\delta} d\Omega < \varepsilon \text{ and } \lim_{\substack{\ell \to 0 \\ m \to \infty}} |\omega(z) - \hat{\omega}_m(z)| = 0.$$

Proof

Choose $0 < \delta < 1$ such that the area of $\Omega - \Omega_\delta$ is less than ε. Let $d = (1-\delta) \min |\zeta|$, $\zeta \in \Gamma$ where $\omega \in W_\Omega$. Then by Theorem 7.3.13, the required result follows.

7.3.15. Discussion

The above theorems discuss the existence of a CVBEM approximation $\hat{\omega}_m(z)$ which converges in measure to $\omega(z)$. That is, for an arbitrarily small $(1-\delta)$-strip inside of Γ, $\hat{\omega}_m(z) \to \omega(z)$ for all $z \in \bar{\Omega}_\delta$ as $m \to \infty$ and $\ell \to 0$. To develop the CVBEM approximation $\hat{\omega}_m(z)$, the defined norm $||x||$ for $x \in W_\Omega$ is used.

To proceed, the $\{f_j\}$ are orthonormalized by the Gram-Schmidt procedure to the set of functions $\{g_j\}$ using the defined inner-product on W_Ω. That is, $g_1 = f_1/||f_1||$, $g_2 = (f_2 - (f_2, g_1)g_1)/||f_2 - (f_2, g_1)g_1||$, and so forth. With respect to $\{g_j\}$,

$$\hat{\omega}_m(z) = \sum_{j=1}^{2(m+2)} \hat{\gamma}_j g_j(z)$$

where the $\hat{\gamma}_j$ are generalized Fourier coefficients to be determined. It is noted that the $g_k(z)$ are finite combinations of the f_j-functions. The value of $||\omega - \hat{\omega}_m||$ is minimized when $\hat{\gamma}_j = (\omega, g_j)$.

By back-substitution, the γ_j corresponding to the $\{f_j\}$ can be evaluated. In this fashion, the CVBEM approximator $\hat{\omega}_m(z)$ is developed for $\omega \in W_\Omega$ and the provided boundary conditions of ϕ defined on Γ_ϕ and ψ defined on Γ_ψ.

Because W_Ω is an inner-product space with the defined inner-product, Bessel's inequality applies.

7.4. The Space $W_\Omega{}^A$

7.4.1. DEFINITION 7.4.1. ($W_\Omega{}^A$)

A subspace of W_Ω are those elements which are analytic over $\bar{\Omega}$. Thus, $\omega \in W_\Omega{}^A$ implies ω is analytic over $\Omega \cup \Gamma$.

7.4.2. Theorem

$W_\Omega{}^A$ is a linear vector space over the field of real numbers.

Proof

Follows from the parent space W_Ω. However, it is noted that ae equality is unnecessary due to $\omega \in W_\Omega{}^A$ implies continuity over $\bar{\Omega}$.

7.4.3. Theorem

$W_\Omega{}^A$ is an inner-product space using the defined inner-product.

Proof

Of interest is showing $(x,x) = 0 \Rightarrow x = 0$. Green's theorem gives

$$\int_\Omega (\phi_x{}^2 + \phi_y{}^2) \, d\Omega = \int_\Gamma \phi \, \frac{\partial \phi}{\partial n} \, d\Gamma + \int_\Omega \phi \nabla^2 \phi \, d\Omega$$

where ϕ_x and ϕ_y are partial derivatives of $\phi(x,y)$ in the x- and y-direction, and $\frac{\partial \phi}{\partial n}$ is a normal derivative along Γ. But $\left| \frac{\partial \phi}{\partial n} \right| = \left| \frac{\partial \psi}{\partial s} \right|$ where s is a tangential coordinate along Γ and the Cauchy-Riemann relations apply. Thus $\nabla^2 \phi = 0$ over Ω due to $\omega = \phi + i\psi$ and $\omega \in W_\Omega{}^A$. Also, $\phi = 0$ on Γ_ϕ and $\frac{\partial \psi}{\partial s} = 0$ on Γ_ψ by assumption.

Thus $\int_\Omega (\phi_x{}^2 + \phi_y{}^2) \, d\Omega = 0$ and $\phi(x,y)$ is constant over Ω. By continuity, $\phi = 0$ over $\bar{\Omega}$. Similarly $\psi = 0$ over $\bar{\Omega}$, and $\omega = 0$.

7.4.4. Discussion

For $\omega \in W_\Omega{}^A$ and $z \in \Omega$, Cauchy's theorem gives immediately that

$$\omega(z) = \frac{1}{2\pi i} \int_\Gamma \frac{\omega(\zeta)d\zeta}{\zeta - z} \, , \quad z \in \Omega \tag{7.4.1}$$

Letting $G_m(\zeta) = \sum_{j=1}^{m} N_j(\zeta)\,\omega_j$ where $\omega_j = \omega(z_j)$, then $\lim\limits_{\substack{m \to \infty \\ \ell \to 0}} G_m(\zeta) = \omega(\zeta)$

and

$$\omega(z) = \lim_{\substack{m \to \infty \\ \ell \to 0}} \frac{1}{2\pi i} \int_\Gamma \frac{G_m(\zeta)d\zeta}{\zeta - z} \, , \quad z \in \Omega \tag{7.4.2}$$

(A detailed proof of this convergence is in Appendix B.)

Thus for $z \in \Omega$,

$$\omega(z) = c_0 + c_{-1}\, z + \sum_{j=1}^{\infty} c_j(z - z_j)\, Ln_j(z - z_j), \quad z \in \Omega \tag{7.4.3}$$

where now c_0 and c_{-1} are also complex constants. It can also be argued that the $c_0 + c_{-1}\, z$ terms can be eliminated entirely when using the infinite series expansion.

Because $\omega(z) = \lim\limits_{\substack{m \to \infty \\ \ell \to 0}} \hat{\omega}_m(z)$ over Ω, then the boundary values of the limiting CVBEM approximator (taking in the limit as $\delta\zeta \to \zeta$ for each $\Omega \in \Gamma$) equal the boundary values of $\omega \in W_\Omega{}^A$.

Writing the $\omega(z)$ function with respect to the Gram-Schmidt orthonormalized functions $\{g_j\}$ of Section 7.3 (with respect to the defined inner-product).

$$\omega(z) = \sum_{j=1}^{\infty} (\omega, g_j)g_j(z), \quad z \in \Omega. \tag{7.4.4}$$

7.4.5. Theorem

The set $\{g_j\}$ is complete.

185

Proof

Suppose $\omega \in W_\Omega A$ and $(\omega, g_j) = 0$ for every j. Then from (7.4.4)

$$\omega(z) = \sum_{j=1}^{\infty} (\omega, g_j) g_j = 0 , \quad z \in \Omega$$

Thus $\omega(z)$ is the zero element of $W_\Omega A$ in that in the limit as $\delta\zeta \to \zeta$, $\phi = 0$ on Γ_ϕ and $\psi = 0$ on Γ_ψ where $\omega = \phi + i\psi$. Thus the set $\{g_j\}$ is complete.

7.4.6. Theorem

Let $\omega \in W_\Omega A$. Then ω satisfies the Dirichlet conditions for generalized Fourier series.

Proof

By assumption, there are a finite number of line segments composing Γ_ϕ and Γ_ψ. Because ω is analytic on Γ, then the boundary condition functions $B(\zeta)$ and $B'(\zeta)$ are both piecewise continuous on Γ.

7.4.7. Discussion: Another Look at W_Ω

By Theorem 7.4.6, the CVBEM will converge to the boundary values where continuous, and to the midpoint value of the discontinuity where discontinuous. Because $\hat{\omega}_m(z)$ is analytic over Ω as $m \to \infty$ (Appendix B), then also $\hat{\omega}_m(z) \to \omega(z)$ as $m \to \infty$. But $\omega(\delta\zeta) \to \omega(\zeta)$ in $L_2(\Gamma)$. Due to $\omega(\delta\zeta)$ being analytic over Ω, we immediately have $\hat{\omega}_m(z)$ approximates $\omega(\delta\zeta)$ which, in turn, approximates $\omega(z)$ arbitrarily close in $L_2(\Gamma)$.

7.5. Applications

7.5.1. Introduction

A FORTRAN CVBEM computer program was prepared based on the least-square boundary fit described in the previous sections. Matrix solution routines to compute coefficients for the CVBEM functions are not needed due to the orthonormal vector technique. The program was prepared to accommodate analytic function equivalents for sources, sinks, flux boundary conditions (i.e. tangential derivatives of the stream function ψ), and dissimilar regions. The program listing is contained in section 7.6.

7.5.2. Nodal Point Placement on Γ

The program operates upon an initial nodal point placement to develop the CVBEM approximation. Then, the user enters (by the CRT) x,y-coordinates for the next node location on Γ and the program computes Bessel's inequality. By spotting the subsequent additional nodal locations on Γ, Bessel's inequality is subsequently minimized and the optimum choice for the next node on Γ is made. In this fashion $\hat{\omega}_m(z) \to \omega(z)$ as $m \to \infty$, (and the nodal spacing decreases).

7.5.3. Potential Flow-Field (Flow-Net) Development

By entering x,y-coordinates, $\hat{\omega}_m(z)$ values are computed and the flow-net can be plotted with respect to the approximation $\hat{\phi}_m(z)$ and $\hat{\psi}_m(z)$ values. Such flow-nets are included in the provided applications.

7.5.4. Approximate Boundary Development

Hromadka (1984) details the "approximate boundary" $\hat{\Gamma}$ technique for CVBEM error evaluation. The contour $\hat{\Gamma}$ represents the location where $\hat{\omega}_m(z)$ achieves the boundary conditions of $\omega(z)$ on Γ. That is, if the provided boundary conditions are level curves of $\omega(z)$ on Γ, then $\hat{\Gamma}$ represents the corresponding level curves of $\hat{\omega}_m(z)$. Hence if the approximate boundary $\hat{\Gamma}$ lies "sufficiently close" to Γ, the analyst can conclude that an adequate approximation has been developed. This error evaluation technique is very useful due to the ease of interpretation. Even beginners can develop highly accurate CVBEM approximations by simply observing the relationship of $\hat{\Gamma}$ to Γ, and adding nodes to Γ where departures are considered unacceptable. In the included example problems, approximate boundaries are developed for each test problem. Further details regarding the approximate boundary technique for error analysis and how it is used with the CVBEM is provided in Appendix C.

7.5.5. Application Problems

Example 7.5.1 (Ideal Fluid Flow Around a Cylindrical Corner)

Ideal fluid flow around a cylindrical corner has the analytic solution of $\omega(z) = z^2 + z^{-2}$. Figure 6.8 depicts the problem geometry and specified boundary conditions. Figures 7.1(a) and 7.1(b) show the error plots in matching boundary values for both the known and unknown boundary conditions. Figure 7.2 shows the CVBEM computed flow net.

Example 7.5.2 (Irregular Domain)

Figure 7.3 shows an irregular two-dimensional cross-section with boundary conditions. The purpose of this example is to show how the approximate boundary is used to evaluate computational error for an irregular section problem. Figure 7.4 shows a very good match between the exact and approximate boundaries.

Example 7.5.3 (Long Shallow Aquifer Groundwater Problem)

A long and shallow unconfined aquifer (see Figure 7.5) is used to compare the results between the dual formulation technique (Frind et al., 1985) and the proposed CVBEM technique. The mean deviation between the exact and approximate boundary is about 0.001^m and 0.2^m for the water table and impervious boundary, respectively. The 0.2^m deviation is based on the 10^{-4} magnitude difference between the exact and approximate boundary. If this magnitude increases to 10^{-2}, the approximate equipotential lines shown on Figure 7.6 approximates those shown on Figures 6 and 8 in Frind et al.'s (1985) paper. The stream lines are not orthogonal to the equipotential lines because of the difference scales in x- and y-directions.

7.6. Computer Program: Two-Dimensional Potential Problems Using Analytic Basis Functions (CVBEM)

7.6.1. Introduction

The CVBEM program consists of two programs—CVBEM1 and CVBEM2. In CVBEM1, boundary conditions are approximated in a "mean-square" error sense and then minimized by the selection of complex coefficients to be associated to each nodal point located on the problem boundary, Γ.

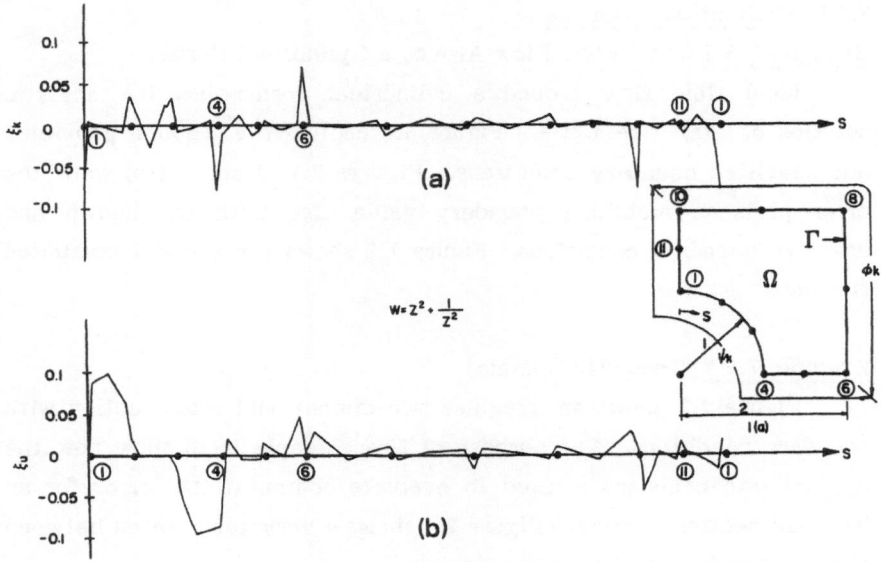

Fig. 7.1. Error Plots for Ideal Fluid Flow Around a Cylindrical Corner.

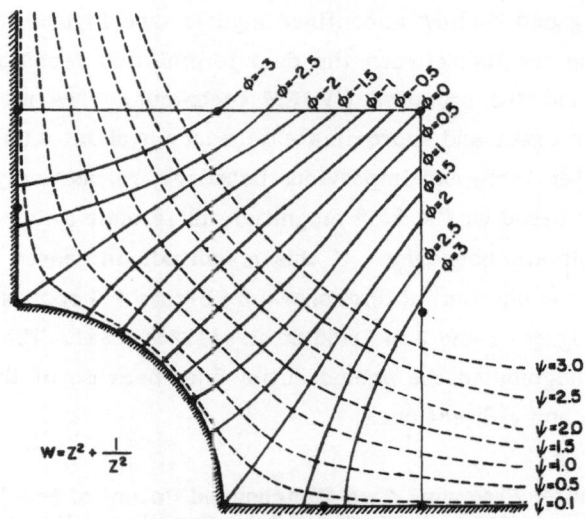

Fig. 7.2. Computed Flow Net for Ideal Fluid Flow Around a Cylindrical Corner.

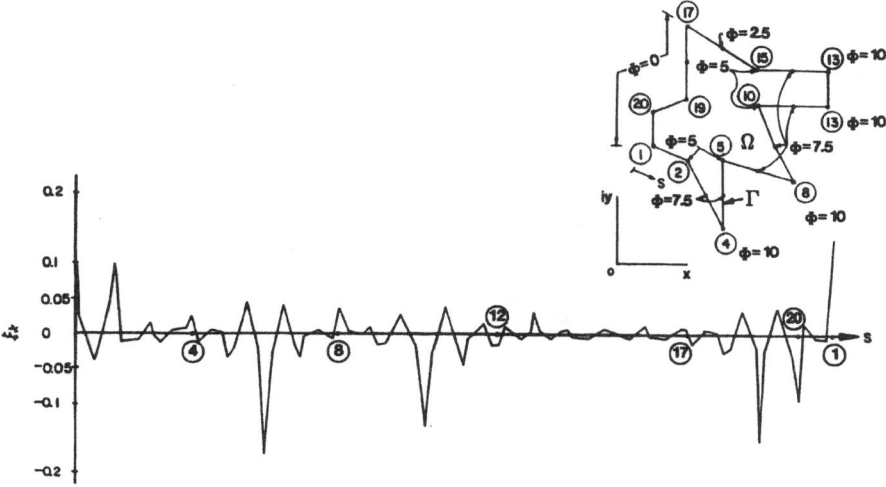

Fig. 7.3.   Error Plot for Irregular Two-Dimensional Cross-section.

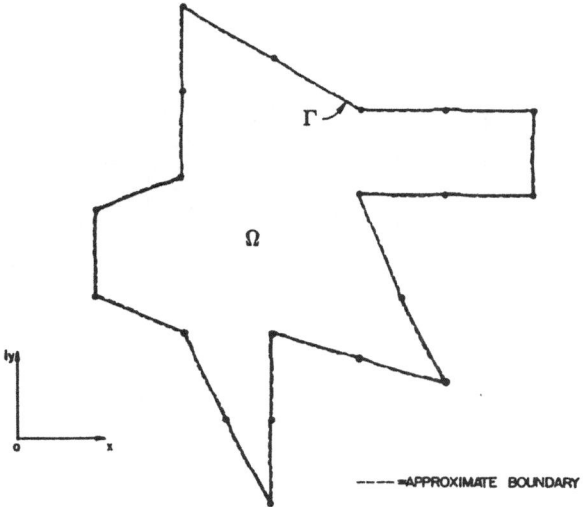

Fig. 7.4.   Approximate Boundary for Irregular Two-Dimensional
Cross-section.

190

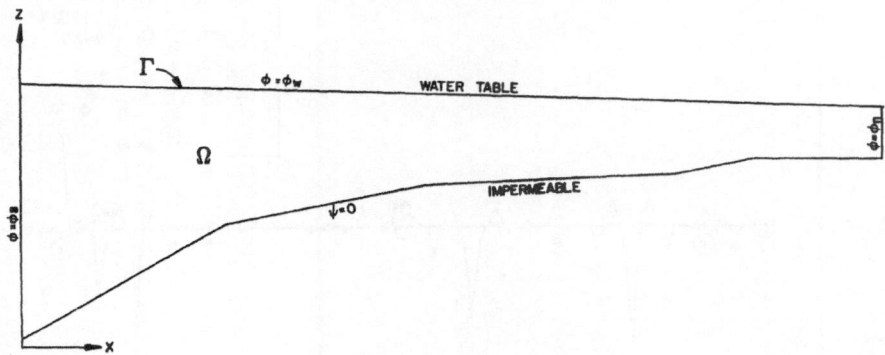

Fig. 7.5.   Boundary Conditions for Shallow Unconfined Aquifer.

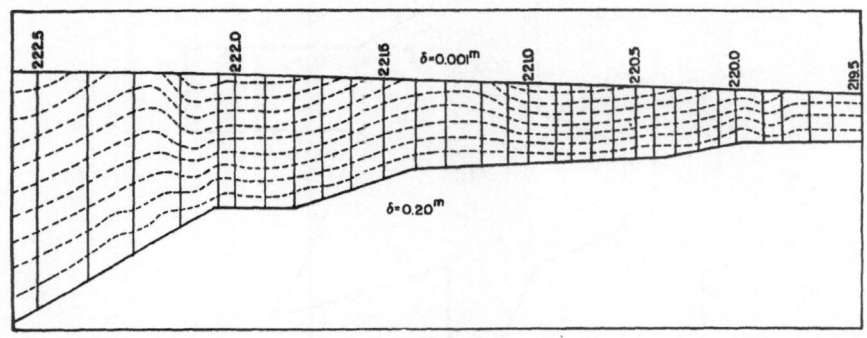

Fig. 7.6.   Computed Flow Net for Shallow Unconfined Aquifer.

The approximate boundary and flow net analysis can be conducted by using CVBEM2.

7.6.2. CVBEM1 Program Listing

Figure 7.7 depicts the simple flow chart for CVBEM1 program.

192

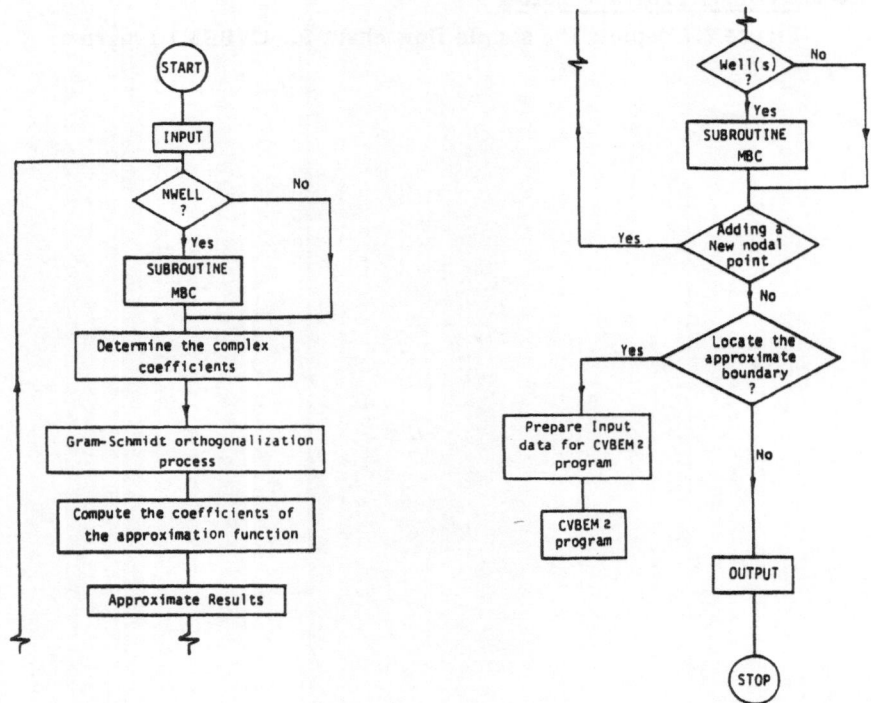

Fig. 7.7.   Simple Flow Chart for Program CVBEM1.

```fortran
$STORAGE:2
C
C   MAIN PROGRAM
C
C   THIS IS A GENERALIZED FOURIER SERIES ANALYSIS
C   WHICH USES CVBEM LINEAR APPROXIMATE FUNCTION TO SOLVE THE
C   LAPLACE EQUATION
C
        IMPLICIT DOUBLE PRECISION(A-H,O-Z)
        COMMON/BLK 1/ X(80),Y(80)
        COMMON/BLK 2/ ANGLE(80),KTYPE(80)
        COMMON/BLK 3/ VALUE(80,2)
        COMMON/BLK 4/ WB(60),B(60)
        COMMON/BLK 5/ G(45,60)
        COMMON/BLK 6/ XK(45,45)
        COMMON/BLK 7/ QX(10),QY(10),Q(10)
C
C   OPEN DATA FILES
C
        NRD=1
        NWT=2
        NWD=3
        NTMR=11
        NTMW=10
        KOK=0
        KADD=0
600     OPEN (UNIT=NRD,FILE='LTWO.DAT',STATUS='OLD')
        IF(KADD.EQ.0) OPEN (UNIT=NWT,FILE='LPT1:',STATUS='OLD')
C
C READ INPUT DATA
C
        PI=3.141592653
        READ(NRD,*) NNOD,NTEST,KODE,KLIN
        IF(KADD.NE.1) GO TO 610
611     WRITE(*,32)
        READ(*,*) NODE
        IF(NODE.GT.NNOD)WRITE(*,929)
        IF(NODE.GT.NNOD)GO TO 611
        DO 620 I=1,NODE
620     READ(NRD,*) X(I),Y(I),KTYPE(I),VALUE(I,1),VALUE(I,2)
     1                ,ANGLE(I)
        K=NODE+1
        IF(K.LE.NNOD)GO TO 621
        X(K)=.5*(X(NODE)+X(1))
        Y(K)=.5*(Y(NODE)+Y(1))
        KTYPE(K)=MIN0(KTYPE(NODE),KTYPE(1))
        ANGLE(K)=ANGLE(NODE)
        DD=DSQRT(((X(K)-X(NODE))**2+(Y(K)-Y(NODE))**2)
        DDD=DSQRT(((X(1)-X(NODE))**2+(Y(1)-Y(NODE))**2)
        DO 647 I=1,2
647     VALUE(K,I)=VALUE(NODE,I)+DD*(VALUE(1,I)-VALUE(NODE,I))/DDD
        KK=K+1
        GO TO 623
621     KK=K+1
        READ(NRD,*) X(KK),Y(KK),KTYPE(KK),VALUE(KK,1),VALUE(KK,2)
     1                ,ANGLE(KK)
        X(K)=.5*(X(NODE)+X(KK))
        Y(K)=.5*(Y(NODE)+Y(KK))
        KTYPE(K)=MIN0(KTYPE(NODE),KTYPE(KK))
        ANGLE(K)=ANGLE(NODE)
```

```
          DD=DSQRT((X(K)-X(NODE))**2+(Y(K)-Y(NODE))**2)
          DDD=DSQRT((X(KK)-X(NODE))**2+(Y(KK)-Y(NODE))**2)
          DO 640 I=1,2
640       VALUE(K,I)=VALUE(NODE,1)+DD*(VALUE(KK,I)-VALUE(NODE,I))/DDD
          KK=KK+1
623       NNOD=NNOD+1
          IF(KK.GT.NNOD)GO TO 643
          GO TO 641
643       READ(NRD,*) NWELL
          IF(NWELL.EQ.0)GO TO 650
          DO 645 I=1,NWELL
645       READ(NRD,*) QX(I),QY(I),Q(I)
          GO TO 650
641       DO 630 I=KK,NNOD
630       READ(NRD,*) X(I),Y(I),KTYPE(I),VALUE(I,1),VALUE(I,2)
      1                 ,ANGLE(I)
          READ(NRD,*) NWELL
          IF(NWELL.EQ.0) GO TO 650
          DO 631 I=1,NWELL
631       READ(NRD,*) QX(I),QY(I),Q(I)
          GO TO 650
610       DO 7 I=1,NNOD
          READ(NRD,*) X(I),Y(I),KTYPE(I),VALUE(I,1),VALUE(I,2)
      1                 ,ANGLE(I)
7         CONTINUE
          READ(NRD,*) NWELL
          IF(NWELL.EQ.0) GO TO 615
          DO 613 I=1,NWELL
613       READ(NRD,*) QX(I),QY(I),Q(I)
615       WRITE(NWT,2)
          DO 9 I=1,NNOD
          IF(KTYPE(I).EQ.1)VALUE(I,2)=0.
          IF(KTYPE(I).EQ.2)VALUE(I,1)=0.
          WRITE(NWT,8) I,X(I),Y(I),KTYPE(I),VALUE(I,1),VALUE(I,2)
      1                 ,ANGLE(I)
9         CONTINUE
          IF(NWELL.EQ.0) GO TO 619
          WRITE(NWT,82)
          DO 85 I=1,NWELL
85        WRITE(NWT,83) I,QX(I),QY(I),Q(I)
619       IF(KOK.EQ.1)GO TO 617
C
C   OUTPUT FORMATS
C
2         FORMAT(//,10X,'*** BOUNDARY NODE DATA ***',
      1//,1X,'NODE',3X,'X(I)',6X,'Y(I)',4X,'KTYPE(I)',5X,
      2'VAUE(I)',9X,'OUTER',/,2X,'NO.',21X,'1=SV;2=SF',3X,
      3'SV',5X,'SF',8X,'NORMAL',/,27X,'3=SV&SF',20X,'ANGLE(I)')
3         FORMAT(2X,2I3,3X,D10.4)
4         FORMAT(//,10X,'*** APPROXIMATE NODAL VALUES AND ERRORS ***',
      1//,6X,'NODE',9X,'STATE',12X,'STREAM',/,
      25X,'NUMBER',6X,'VARIABLE',10X,'FUNCTION',12X,'ERROR',/)
45        FORMAT(//,10X,'*** APPROXIMATE POINT VALUES AND ERRORS --',
      1' FUNCTION FORM ***',//,5X,'POINT',9X,'STATE',12X,'STREAM',/,
      25X,'NUMBER',6X,'VARIABLE',10X,'FUNCTION',12X,'ERROR',/)
5         FORMAT(3X,I5,3(8X,D10.4))
8         FORMAT(1X,I3,2X,F8.3,2X,F8.3,5X,I2,4X,F7.2,1X,F7.2,5X,F6.2)
21        FORMAT(///,10X,'*** EVALUATION POINT DATA ***',
      1//,1X,'POINT',2X,'X(I)',6X,'Y(I)',4X,'KTYPE(I)',5X,
      2'VALUE(I)',/,2X,'NO.',21X,'1=SV;2=SF',3X,'SV',5X,'SF',/,
```

```
       327X,'3=SV&SF')
22     FORMAT(10(1X,D10.4))
23     FORMAT(/,10X,'*** NODE NUMBER #',I3,2X,'(REAL,IMAGINARY)')
24     FORMAT(/,10X,'*** NODAL POINT VECTOR EXPANSION, F(I) ***',/)
25     FORMAT(/,10X,'*** ORTHOGONAL VECTOR EXPANSION, G(I) ***',/)
26     FORMAT(/,10X,'*** ORTHOGONAL TEST (G(I),G(I)) ***',//,
     1'      I   J   (G(I),G(J))')
28     FORMAT(/,10X,'*** EVALUATION COEFFICIENTS,',
     1' (G(I),B(I))/(G(I),G(I)) ***',/)
29     FORMAT(/,10X,'*** BACKSUBSTITUTION COEFFICIENTS ***',/)
31     FORMAT(/,2X,'ENTER A [1] FOR ADDING AN ADDITIONAL NODAL POINT')
32     FORMAT(/,2X,'ENTER THE NODE NUMBER THAT',/,
     12X,'THE ADDITIONAL NODAL POINT WILL FOLLOWED')
33     FORMAT(/,2X,'ENTER THE X- AND Y- COORDIATES',/,
     12X,'FOR THE ADDITIONAL NODAL POINT')
34     FORMAT(/,2X,'THE ADDITIONAL NODAL POINT (',F8.4,',',F8.4,
     1') ',//,'  HAS THE NORM EQUAL TO ',D10.4)
35     FORMAT(/,2X,'ENTER A [1] TO ACCEPT THIS ADDITIONAL NODAL POINT')
36     FORMAT(/,2X,'*** BESSEL INEQUALITY ***',/,2X,D10.4,
     1' >= ',D10.4,' AND THE DIFFERENCE IS ',D10.4,/,120('-'),/)
38     FORMAT(2(1X,F10.4),1X,I2,3(1X,F10.4))
71     FORMAT(2X,'ENTER A [1] TO EVALUATE THE STATE VARIABLE AND',
     1/,'   STREAM FUNCTION FOR A GIVEN POINT')
72     FORMAT(2X,'===> EXECUTE PROGRAM "CCYLTWO1" TO EVALUATE',/,
     1'    STATE VARIABLE AND STREAM FUNCTION FOR A GIVEN POINT')
81     FORMAT(1X,I3,2X,F8.3,2X,F8.3,5X,I2,4X,F7.2,1X,F7.2)
82     FORMAT(/,2X,'*** LOCATION AND STRENGTH OF WELL(S) ***',/)
83     FORMAT(1X,I3,2(2X,F8.3),7X,F10.5)
99     FORMAT(/,120('='))
919    FORMAT(2X,'DATA POINT HAS BEEN USED...TRY ANOTHER POINT')
929    FORMAT(2X,'NUMBER EXCEEDS TOTAL NODE NUMBER...TRY AGAIN')
C
C  CALCULATE X-COORINATE, Y-COORIDNATE, AND BOUNDARY VALUES
C  FOR EVALUATION POINTS
C
650    K=NNOD
307    DO 300 I=1,NNOD
       IP1=I+1
       IF(IP1.GT.NNOD)IP1=1
       X1=X(I)
       Y1=Y(I)
       X2=X(IP1)
       Y2=Y(IP1)
       DX=X2-X1
       DY=Y2-Y1
       R11=VALUE(I,1)
       R12=VALUE(I,2)
C..MODIFY THE EFFECT BY THE WELL(S)
       IF(NWELL.NE.0) CALL MBC(NWELL,R11,R12,X1,Y1,1)
       R21=VALUE(IP1,1)
       R22=VALUE(IP1,2)
C..MODIFY THE EFFECT BY THE WELL(S)
       IF(NWELL.NE.0) CALL MBC(NWELL,R21,R22,X2,Y2,1)
       DR1=R21-R11
       DR2=R22-R12
       DO 300 J=1,NTEST
       K=K+1
       P=1./(2.*(NTEST+1))
       IF(NTEST.EQ.1) RATIO=0.5
       IF(NTEST.GT.1) RATIO=P+(1.-2.*P)*FLOAT(J-1)/(FLOAT(NTEST)-1.)
```

```
        X(K)=X1+DX*RATIO
        Y(K)=Y1+DY*RATIO
        ANGLE(K)=ANGLE(I)
        VALUE(K,1)=R11+DR1*RATIO
        VALUE(K,2)=R12+DR2*RATIO
        KTYPE(K)=MINO(KTYPE(I),KTYPE(IP1))
300     CONTINUE
        NTOT=NNOD*NTEST
        NTOTL=2*NNOD
        NNODP=NNOD
        IF(KLIN.EQ.1)NTOTL=2*NNOD+4
        IF(KLIN.EQ.1)NNODP=NNCD+2
617     IF(KADD.EQ.0 .OR. KOK.EQ.1)WRITE(NWT,21)
        SAREA=0.
        DO 310 KK=1,NTOT
        I=KK+NNOD
        IF(KTYPE(I).EQ.1)VALUE(I,2)=0.
        IF(KTYPE(I).EQ.2)VALUE(I,1)=0.
        SAREA=SAREA+VALUE(I,1)**2+VALUE(I,2)**2
        IF(KADD.EQ.0 .OR. KOK.EQ.1)WRITE(NWT,81)KK,X(I),Y(I),KTYPE(I),
     1                            VALUE(I,1),VALUE(I,2)
        IF(KOK.EQ.1)GO TO 310
        ANGLE(I)=ANGLE(I)*PI/180.
310     CONTINUE
C
C   DETERMINE THE COEFFICIENTS OF ALPHA'S AND BETA'S FOR
C   COMPLEX VARIABLE (Z-Z(J))*LN(Z-Z(J)) ARITHMETIC
C
        IF(KOK.EQ.1)GO TO 660
        IF(KODE.EQ.1)WRITE(NWT,99)
        IF(KODE.EQ.1)WRITE(NWT,24)
        DO 320 I=1,NNODP
        II=2*I
        IIM1=II-1
        XX=X(I)
        YY=Y(I)
        DO 330 KK=1,NTOT
        J=KK+NNOD
        XJ=X(J)
        YJ=Y(J)
        IF(I.EQ.NNOD+1)GO TO 339
        IF(I.EQ.NNOD+2)GO TO 331
        A1=XJ-XX
        B1=YJ-YY
        RJ=DSQRT(A1*A1+B1*B1)
        D=DLOG(RJ)
        CALL CAUCH5(A1,B1,ANG)
        TH=ANG-ANGLE(I)*PI/180.
        IF(TH.LT.0.)TH=TH+2.*PI
        ALPHA=A1*D-TH*B1
        BETA=B1*D+TH*A1
        GO TO 337
331     ALPHA=XJ
        BETA=YJ
        GO TO 337
339     ALPHA=1.
        BETA=0.
337     GO TO(335,345,335)KTYPE(J)
335     G(IIM1,KK)=ALPHA
        G(II,KK)=-1.*BETA
```

```
      GO TO 330
345   G(IIM1,KK)=BETA
      G(II,KK)=ALPHA
330   CONTINUE
      IF(KODE.NE.1)GO TO 320
      WRITE(NWT,23)I
      WRITE(NWT,22)(G(IIM1,K),K=1,NTOT)
      WRITE(NWT,22)(G(II,K),K=1,NTOT)
320   CONTINUE
C
C  USE THE GRAM-SCHMIDT ORTHOGONALIZATION PROCESS
C  TO DETERMINE SERIES OF ORTHOGONAL VECTORS
C
      DO 20 I=2,NTOTL
      DO 20 KK=2,I
      SUM1=0.
      SUM2=0.
      DO 30 J=1,NTOT
      SUM1=SUM1+G(I,J)*G(KK-1,J)
      SUM2=SUM2+G(KK-1,J)*G(KK-1,J)
30    CONTINUE
      XXK=-1.*SUM1/SUM2
      XK(I,KK-1)=XXK
      DO 40 J=1,NTOT
      G(I,J)=G(I,J)+XXK*G(KK-1,J)
40    CONTINUE
20    CONTINUE
      IF(KODE.NE.1)GO TO 55
      WRITE(NWT,25)
      DO 47 I=1,NTOTL
47    WRITE(NWT,22)(G(I,J),J=1,NTOT)
C..CHECK ORTHOGONALITY OF VECTORS G(I)
      WRITE(NWT,26)
55    DO 50 I=1,NTOTL
      IP1=I+1
      IF(I.EQ.NTOTL)GO TO 80
      DO 70 K=IP1,NTOTL
      SUM=0.
      DO 60 J=1,NTOT
60    SUM=SUM+G(I,J)*G(K,J)
      IF(KODE.EQ.1)WRITE(NWT,3)I,K,SUM
      IF(KODE.NE.1 .AND. ABS(SUM).GT..00001)WRITE(NWT,3)I,K,SUM
70    CONTINUE
50    CONTINUE
80    CONTINUE
C..COMPUTE THE COEFFICIENTS OF B(I)=(W,G(I))/(G(I),G(I))
      IF(KODE.EQ.1)WRITE(NWT,28)
      SUM=0.
      DO 120 I=1,NTOTL
      BK1=0.
      BK2=0.
      DO 130 KK=1,NTOT
      J=KK+NNOD
      IF(KTYPE(J).EQ.1 .OR. KTYPE(J).EQ.3)BK1=BK1+VALUE(J,1)*G(I,KK)
      IF(KTYPE(J).EQ.2)BK1=BK1+VALUE(J,2)*G(I,KK)
      BK2=BK2+G(I,KK)*G(I,KK)
130   CONTINUE
C..COMPUTE THE NORM OF THE GENERALIZED FOURIER COEFFICIENTS
      B(I)=BK1/BK2
      SUM=SUM+BK1*BK1/BK2
```

```
120     CONTINUE
        IF(KODE.EQ.1)WRITE(NWT,22)(B(I),I=1,NTOTL)
        IF(KADD.NE.1)GO TO 660
        IJ=NODE+1
        WRITE(*,34)X(IJ),Y(IJ),SUM
        WRITE(NWT,34)X(IJ),Y(IJ),SUM
        WRITE(*,35)
        READ(*,*)KOK
        IF(KOK.NE.1)KOK=0
        IF(KOK.NE.1)GO TO 680
        WRITE(NWT,99)
        WRITE(NWT,34)X(IJ),Y(IJ),SUM
        GO TO 615
C..COMPUTE THE COEFFICIENTS OF THE APPROXIMATE FUNCTIONS
660     DO 200 I=NTOTL,1,-1
        IF(I.EQ.NTOTL)XK(NTOTL,I)=B(NTOTL)
        IF(I.NE.NTOTL)XK(NTOTL,I)=XK(NTOTL,I)*B(NTOTL)+B(I)
200     CONTINUE
        NTOT1=NTOTL-1
        DO 210 I=NTOT1,1,-1
        DO 210 J=I,1,-1
        IF(I.EQ.J)GO TO 210
        IF(I.NE.J)XK(NTOTL,J)=XK(NTOTL,I)*XK(I,J)+XK(NTOTL,J)
210     CONTINUE
        IF(KODE.EQ.1)WRITE(NWT,29)
        IF(KODE.EQ.1)WRITE(NWT,22)(XK(NTOTL,I),I=1,NTOTL)
        DO 110 I=1,NTOT
        B(I)=0.
        WB(I)=0.
110     CONTINUE
C..APPROXIMATE THE EVALUATION POINT -- FUNCTION FORM
        WRITE(NWT,99)
        WRITE(NWT,45)
        DO 480 I=1,NTOT
        II=I+NNOD
        XX=X(II)
        YY=Y(II)
        XRE=0.
        XIM=0.
        DO 470 J=1,NNODP
        JJ=J*2
        JJM1=JJ-1
        IF(J.EQ.NNOD+1)GO TO 471
        IF(J.EQ.NNOD+2)GO TO 473
        XJ=X(J)
        YJ=Y(J)
        A1=XX-XJ
        B1=YY-YJ
        RJ=DSQRT(A1*A1+B1*B1)
        D=DLOG(RJ)
        CALL CAUCH5(A1,B1,ANG)
        TH=ANG-ANGLE(J)*PI/180.
        IF(TH.LT.0.)TH=2.*PI+TH
        ALPHA=A1*D-TH*B1
        BETA=B1*D+TH*A1
        GO TO 475
471     ALPHA=1.
        BETA=0.
        GO TO 475
473     ALPHA=XX
```

```
        BETA=YY
475     XRE=XRE+XK(NTOTL,JJM1)*ALPHA-XK(NTOTL,JJ)*BETA
        XIM=XIM+XK(NTOTL,JJM1)*BETA+XK(NTOTL,JJ)*ALPHA
470     CONTINUE
        XD=XRE-VALUE(II,1)
        IF(KTYPE(II).EQ.2)XD=XIM-VALUE(II,2)
C..MODIFY THE EFFECT BY THE WELL(S)
        IF(NWELL.NE.0) CALL MBC(NWELL,XRE,XIM,XX,YY,2)
        WRITE(NWT,5)I,XRE,XIM,XD
480     CONTINUE
C..APPROXIMATE THE NODAL VALUES -- FUNCTION FORM
290     WRITE(NWT,99)
        WRITE(NWT,4)
        DO 280 I=1,NNOD
        XX=X(I)
        YY=Y(I)
        XRE=0.
        XIM=0.
        DO 270 J=1,NNODP
        IF(I.EQ.J)GO TO 270
        JJ=J*2
        JJM1=JJ-1
        IF(J.EQ.NNOD+1)GO TO 271
        IF(J.EQ.NNOD+2)GO TO 273
        XJ=X(J)
        YJ=Y(J)
        A1=XX-XJ
        B1=YY-YJ
        RJ=DSQRT(A1*A1+B1*B1)
        D=DLOG(RJ)
        CALL CAUCH5(A1,B1,ANG)
        TH=ANG-ANGLE(J)*PI/180.
        IF(TH.LT.0.)TH=2.*PI+TH
        ALPHA=A1*D-TH*B1
        BETA=B1*D+TH*A1
        GO TO 275
271     ALPHA=1.
        BETA=0.
        GO TO 275
273     ALPHA=XX
        BETA=YY
275     XRE=XRE+XK(NTOTL,JJM1)*ALPHA-XK(NTOTL,JJ)*BETA
        XIM=XIM+XK(NTOTL,JJM1)*BETA+XK(NTOTL,JJ)*ALPHA
270     CONTINUE
C..APPROXIMATE THE NODAL VALUES -- FUNCTION FORM
        IF(NWELL.NE.0) CALL MBC(NWELL,XRE,XIM,XX,YY,2)
        XD=XRE-VALUE(I,1)
        IF(KTYPE(I).EQ.2)XD=XIM-VALUE(I,2)
        WRITE(NWT,5)I,XRE,XIM,XD
280     CONTINUE
        WRITE(NWT,99)
C..BESSEL'S INEQUALITY
        DIFF=SAREA-SUM
        IF(ABS(DIFF).LT.0.00001)DIFF=0.
        WRITE(NWT,36)SAREA,SUM,DIFF
        ERROR=DIFF/SAREA
        WRITE(*,31)
        READ(*,*)KADD
        IF(KADD.NE.1)KADD=0
        IF(KADD.NE.1)GO TO 260
```

```
        IF(KOK.EQ.1)GO TO 690
680     CLOSE(UNIT=NRD,STATUS='KEEP')
        GO TO 600
690     CLOSE(UNIT=NRD,STATUS='KEEP')
        OPEN (UNIT=NRD,FILE='LTWO.DAT',STATUS='OLD')
        WRITE(NRD,*) NNOD,NTEST,KODE,KLIN
        DO 605 I=1,NNOD
        IF(KTYPE(I).EQ.1)VALUE(I,2)=0.
        IF(KTYPE(I).EQ.2)VALUE(I,1)=0.
        WRITE(NRD,*)X(I),Y(I),KTYPE(I)
        WRITE(NRD,*)VALUE(I,1),VALUE(I,2),ANGLE(I)
605     CONTINUE
        WRITE(NRD,*)NWELL
        IF(NWELL.EQ.0) GO TO 607
607     DO 609 J=1,NWELL
609     WRITE(NRD,*)QX(J),QY(J),Q(J)
        CLOSE (UNIT=NRD,STATUS='KEEP')
        KOK=0
        GO TO 600
260     WRITE(*,71)
        READ(*,*) KEVA
        IF(KEVA.NE.1) GO TO 700
        OPEN (UNIT=NWD,FILE='LTWO1.DAT',STATUS='OLD')
        WRITE(NWD,*) NNOD,NNODP,NTOTL
        DO 710 I=1,NNOD
710     WRITE(NWD,*) X(I),Y(I),ANGLE(I)
        WRITE(NWD,*) (XK(NTOTL,I),I=1,NTOTL)
        WRITE(NWD,*) NWELL
        IF(NWELL.EQ.0) GO TO 720
        DO 730 I=1,NWELL
730     WRITE(NWD,*) QX(I),QY(I),Q(I)
720     CLOSE (UNIT=NWD,STATUS='KEEP')
        WRITE(*,72)
700     CLOSE(UNIT=NRD,STATUS='KEEP')
        CLOSE(UNIT=NWT,STATUS='KEEP')
        STOP
        END
```

```
$STORAGE:2
      SUBROUTINE MBC(NWELL,XRE,XIM,X,Y,KODE)
      IMPLICIT DOUBLE PRECISION (A-H,O-Z)
      COMMON/BLK 7/ QX(10),QY(10),Q(10)
C
C  THIS SUBROUTINE MODIFIES THE BOUNDARY CONDITIONS AND THE
C  APPROXIMATE SOLUTION ACCORDING TO THE WELL(S) EFFECT.
C
      TPI=6.283185307
      SGN=1.
      IF(KODE.NE.1)SGN=-1.
      DO 10 J=1,NWELL
      XX=X-QX(J)
      YY=Y-QY(J)
      RJ=DSQRT(XX*XX+YY*YY)
      D=DLOG(RJ)
      XRE=XRE-SGN*(Q(J)*D)/TPI
      CALL CAUCH5(XX,YY,B)
      XIM=XIM-SGN*(Q(J)*B)/TPI
10    CONTINUE
      RETURN
      END
C
```

```
$STORAGE:2
C----------------------------------------------------------
C       SUBROUTINE CAUCH5
C----------------------------------------------------------
        SUBROUTINE CAUCH5(X,Y,ANGLE)
        IMPLICIT DOUBLE PRECISION(A-H,O-Z)
C
C  THIS SUBROUTINE DETERMINES THE POSITIVE ANGLE
C  OF COMPLEX POINT Z WITH RESPECT TO THE ORIGIN
C
        PI=3.141592653
        IF(X.EQ.0.  .AND.  Y.GT.0.)ANGLE=.5*PI
        IF(X.EQ.0.  .AND.  Y.LT.0.)ANGLE=1.5*PI
        IF(X.GT.0.  .AND.  Y.GE.0.)ANGLE=DATAN(Y/X)
        IF(X.LT.0.  .AND.  Y.GE.0.)ANGLE=PI-DATAN(-Y/X)
        IF(X.LT.0.  .AND.  Y.LT.0.)ANGLE=PI+DATAN(Y/X)
        IF(X.GT.0.  .AND.  Y.LT.0.)ANGLE=2.*PI-DATAN(-Y/X)
        IF(X.EQ.0.  .AND.  Y.EQ.0.)ANGLE=0.
        RETURN
        END
C
```

7.6.3. Input Variable Description for CVBEM1

The input file has the following form:

Line	Variables	
1	NNOD,NTEST,KODE,KLIN	
2	X(I),Y(I),KTYPE(I),VALUE(I,1), VALUE(I,2),ANGLE(I)	⎫
.		⎬ NNOD
.		⎭
NNOD+1		
NNOD+2	NWELL	
NNOD+3	QX(I),QY(I),Q(I)	⎫
NNOD+2+NWELL		⎬ NWELL ⎭

where

NNOD is the total node number.

NTEST is the number of evaluation points per element.

$KODE = \begin{cases} 0, & \text{Summary output} \\ 1, & \text{Detail output} \end{cases}$

$KLIN = \begin{cases} 0, & \text{excludes the linear part of the approximation function} \\ 1, & \text{includes the linear part of the approximation function} \end{cases}$

X(I) is the x-coordinate

Y(I) is the y-coordinate

$KTYPE(I) = \begin{cases} 1, & \text{state variable specified for Node I} \\ 2, & \text{stream function specified for Node I} \\ 3, & \text{state variable and stream function specified for Node I} \\ 4, & \text{efflux boundary condition} \end{cases}$

VALUE(I,1) is the value of state variable for Node I

VALUE(I,2) is the value of stream function for Node I

ANGLE(I)   is the outer normal for Node I

NWELL   is the number of sink or source terms

QX(I)   is the x-coordinate for the Ith sink or
           source term

QY(I)   is the y-coordinate for the Ith sink or
           source term

Q(I)   is the strength of the sink (+) or source (-) term

Note that all the source and sinks have a branch cut parallel to the real x-axis.

7.6.4. CVBEM2 Program Listing

Figure 7.8 depicts the simple flow chart of the CVBEM2 program and the listing of the program is included in this section.

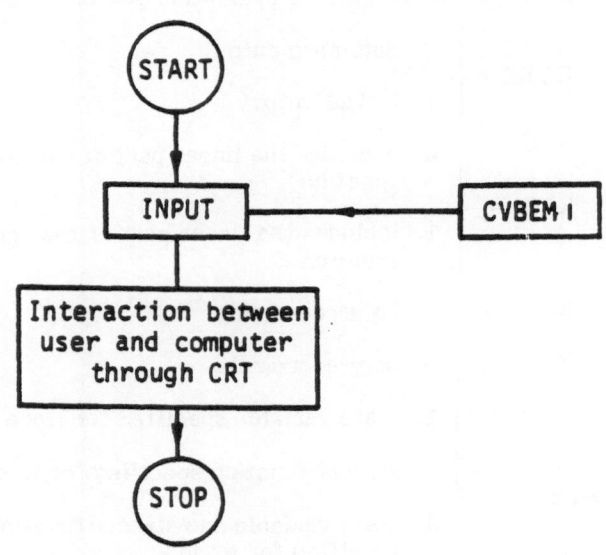

Fig. 7.8. Simple Flow Chart for Program CVBEM2.

```
$STORAGE:2
C
C  MAIN PROGRAM
C
C  THIS PROGRAM EVALUATES THE STATE VARIABLE AND STREAM FUNCTION
C  FOR A GIVEN NODAL POINT
C
       IMPLICIT DOUBLE PRECISION(A-H,O-Z)
       COMMON/BLK 1/ X(50),Y(50)
       COMMON/BLK 2/ ANGLE(50),XK(50)
       COMMON/BLK 7/ QX(10),QY(10),Q(10)
C
C  OPEN DATA FILES
C
       NRD=1
       NWD=2
       NTR=0
       NTW=0
       OPEN (UNIT=NRD,FILE='LTWO1.DAT',STATUS='OLD')
C
C READ INPUT DATA
C
       PI=3.141592653
       READ(NRD,*) NNOD,NNODP,NTOTL
       DO 10 I=1,NNOD
       READ(NRD,*) X(I),Y(I),ANGLE(I)
10     CONTINUE
       READ(NRD,*) (XK(I),I=1,NTOTL)
       READ(NRD,*) NWELL
       IF(NWELL.EQ.0) GO TO 15
       DO 17 I=1,NWELL
17     READ(NRD,*) QX(I),QY(I),Q(I)
15     CLOSE (UNIT=NRD,STATUS='KEEP')
       DO 18 I=1,NNOD
       ANGLE(I)=ANGLE(I)*PI/180.
18     CONTINUE
C
C  OUTPUT FORMATS
C
4      FORMAT(/,2X,'*** INVALID DATA ENTRY...TRY AGAIN ***',/)
5      FORMAT(3X,F10.4,2X,F10.4,2X,2(2X,F10.3))
6      FORMAT(10X,'*** APPROXIMATE SOLUTION ***',/)
7      FORMAT(2X,'EVALUATE THE STATE VARIABLE AND STREAM FUNCTION',
      1' FOR A GIVEN POINT',/,2X,'ENTER THE X-COORDINATE')
8      FORMAT(2X,'ENTER THE Y-COORDINATE')
9      FORMAT(/,7X,'X-',10X,'Y-',10X,'STATE',6X,'STREAM',
      1/,3X,'COORDINATE',2X,'COORDINATE',4X,'VARIABLE',4X,
      2'FUNCTION',/)
11     FORMAT(/,2X,'ENTER THE OPTION :',/,3X,
      1'[1] FOR ANOTHER POINT',/,3X,'[2] FOR ACCEPTING THE'
      2,' CURRENT POINT AND STARTING A NEW POINT',/,3X,
      3'[3] FOR ACCEPTING THE CURRENT POINT AND TERMINATING'
      4,' THE PROCESS')
12     FORMAT(/,2X,'*** THE NODAL POINT HAS BEEN USED...TRY AGAIN'
      1,' ***',/)
C
C  APPROXIMATE THE STATE VARIABLE AND STREAM FUNCTIONS
C
       OPEN (UNIT=NWD,FILE='LPT1:',STATUS='NEW')
       WRITE(NWD,6)
```

```
          WRITE(NWD,9)
300       WRITE(NTW,7)
          READ(NTR,*) XX
          WRITE(NTW,8)
          READ(NTR,*) YY
          DO 100 I=1,NNOD
          IF(XX.EQ.X(I) .AND. YY.EQ.Y(I))GO TO 110
100       CONTINUE
          GO TO 120
110       WRITE(NTW,12)
          GO TO 300
120       XRE=0.
          XIM=0.
          DO 270 J=1,NNODP
          JJ=J*2
          JJM1=JJ-1
          IF(J.EQ.NNOD+1)GO TO 271
          IF(J.EQ.NNOD+2)GO TO 273
          A1=XX-X(J)
          B1=YY-Y(J)
          RJ=DSQRT(A1*A1+B1*B1)
          D=DLOG(RJ)
          CALL CAUCH5(A1,B1,ANG)
          TH=ANG-ANGLE(J)*PI/180.
          IF(TH.LT.0.)TH=2.*PI+TH
          ALPHA=A1*D-TH*B1
          BETA=B1*D+TH*A1
          GO TO 275
271       ALPHA=1.
          BETA=0.
          GO TO 275
273       ALPHA=XX
          BETA=YY
275       XRE=XRE+XK(JJM1)*ALPHA-XK(JJ)*BETA
          XIM=XIM+XK(JJM1)*BETA+XK(JJ)*ALPHA
270       CONTINUE
C..MODIFY THE EFFECT BY THE WELL(S)
          IF(NWELL.NE.0) CALL MBC(NWELL,XRE,XIM,XX,YY,2)
          WRITE(NTW,9)
          WRITE(NTW,5)XX,YY,XRE,XIM
330       WRITE(NTW,11)
          READ(NTR,*) KODE
          IF(KODE.EQ.1) GO TO 300
          IF(KODE.EQ.2) GO TO 310
          IF(KODE.EQ.3) GO TO 320
          WRITE(NTW,4)
          GO TO 330
310       WRITE(NWD,5)XX,YY,XRE,XIM
          GO TO 300
320       WRITE(NWD,5)XX,YY,XRE,XIM
          CLOSE(UNIT=NWD,STATUS='KEEP')
          STOP
          END
```

```
$STORAGE:2
      SUBROUTINE MBC(NWELL,XRE,XIM,X,Y,KODE)
      IMPLICIT DOUBLE PRECISION (A-H,O-Z)
      COMMON/BLK 7/ QX(10),QY(10),Q(10)
C
C  THIS SUBROUTINE MODIFIES THE BOUNDARY CONDITIONS AND THE
C  APPROXIMATE SOLUTION ACCORDING TO THE WELL(S) EFFECT.
C
      TPI=6.283185307
      SGN=1.
      IF(KODE.NE.1)SGN=-1.
      DO 10 J=1,NWELL
      XX=X-QX(J)
      YY=Y-QY(J)
      RJ=DSQRT(XX*XX+YY*YY)
      D=DLOG(RJ)
      XRE=XRE-SGN*(Q(J)*D)/TPI
      CALL CAUCH5(XX,YY,B)
      XIM=XIM-SGN*(Q(J)*B)/TPI
10    CONTINUE
      RETURN
      END
C
```

```
$STORAGE:2
C------------------------------------------------------------
C      SUBROUTINE CAUCH5
C------------------------------------------------------------
       SUBROUTINE CAUCH5(X,Y,ANGLE)
       IMPLICIT DOUBLE PRECISION(A-H,O-Z)
C
C   THIS SUBROUTINE DETERMINES THE POSITIVE ANGLE
C   OF COMPLEX POINT Z WITH RESPECT TO THE ORIGIN
C
       PI=3.141592653
       IF(X.EQ.0. .AND. Y.GT.0.)ANGLE=.5*PI
       IF(X.EQ.0. .AND. Y.LT.0.)ANGLE=1.5*PI
       IF(X.GT.0. .AND. Y.GE.0.)ANGLE=DATAN(Y/X)
       IF(X.LT.0. .AND. Y.GE.0.)ANGLE=PI-DATAN(-Y/X)
       IF(X.LT.0. .AND. Y.LT.0.)ANGLE=PI+DATAN(Y/X)
       IF(X.GT.0. .AND. Y.LT.0.)ANGLE=2.*PI-DATAN(-Y/X)
       IF(X.EQ.0. .AND. Y.EQ.0.)ANGLE=0.
       RETURN
       END
C
```

Example 7.6.1. (Example Input–Output Data)

Approximate the analytical function $\omega = e^z$ on a π-square region. Figure 7.9 shows the nodal points placement and depicts the boundary conditions.

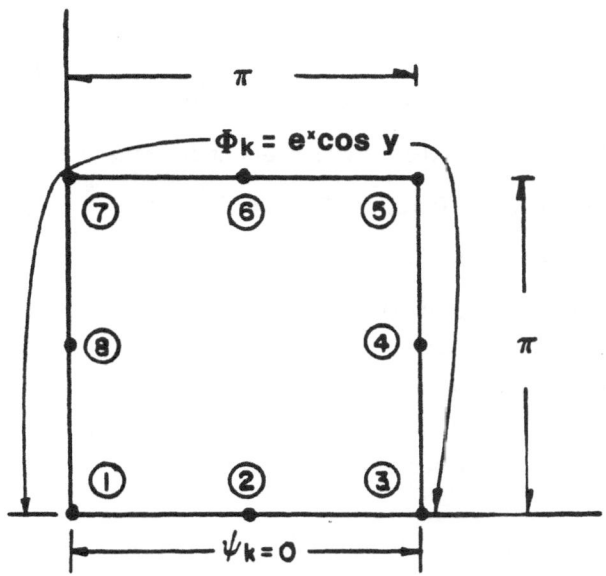

Fig. 7.9. Example Input-output Problem ($\omega = e^z$) for Program CVBEM1.

The input date file is stored in "LTWO.DAT" which consists of:

8	5	1	1		
0	0	3	1	0	270
1.570796	0	2	0	0	270
3.141592	0	3	23.14069	0	0
3.141592	1.570796	1	0	0	0
3.141592	3.141592	1	-23.14069	0	90
1.570796	3.141592	1	-4.81048	0	90
0	3.141592	1	-1	0	180
0	1.570796	1	0	0	180
0					

END OF FILE

Output File Description

The output file is stored in "LTWO.ANS" which has the following form:

(a) Boundary Node Data

(b) Evaluation Point Data

(c) Nodal Point Vector Expansion (Optional)
The vector is filled by the real or imaginary part of the approximate function $(C_{m+1} + C_{m+2} z + \sum_{j=1}^{m} (z-z_j)(Ln(z-z_j))$ where m is the total node number) according to the boundary conditions

(d) Orthogonal Vector Expansion (Optional)
Using the Gram–Schmidt orthonormalization process to determine series of orthogonal vectors of part (c)

(e) Orthogonal Test (Optional)

(f) Evaluation Coefficients (Optional)
Coefficients for orthogonal vector of part (d)

(g) Back Substitution Coefficients (Optional)
Coefficients for original vector of part (c)

(h) Approximation Point Values and Errors
The Error is defined as the difference between the approximation value and the specified boundary condition

(i) Approximation Nodal Values and Errors

(j) Bessel's Inequality
The left hand side of the inequality sign is the dot product of the boundary condition vector and the right hand side is the dot product of the normalized Fourier coefficients

The output file is included hereafter.

*** BOUNDARY NODE DATA ***

NODE NO.	X(I)	Y(I)	KTYPE(I) 1=SV;2=SF 3=SV&SF	VALUE(I) SV	SF	OUTER NORMAL ANGLE(I)
1	.000	.000	3	1.00	.00	270.00
2	1.571	.000	2	.00	.00	270.00
3	3.142	.000	3	23.14	.00	.00
4	3.142	1.571	1	.00	.00	.00
5	3.142	3.142	1	-23.14	.00	90.00
6	1.571	3.142	1	-4.81	.00	90.00
7	.000	3.143	1	-1.00	.00	180.00
8	.000	1.571	1	.00	.00	180.00

*** EVALUATION POINT DATA ***

POINT NO.	X(I)	Y(I)	KTYPE(I) 1=SV;2=SF 3=SV&SF	VALUE(I) SV	SF
1	.131	.000	2	.00	.00
2	.458	.000	2	.00	.00
3	.785	.000	2	.00	.00
4	1.113	.000	2	.00	.00
5	1.440	.000	2	.00	.00
6	1.702	.000	2	.00	.00
7	2.029	.000	2	.00	.00
8	2.356	.000	2	.00	.00
9	2.683	.000	2	.00	.00
10	3.011	.000	2	.00	.00
11	3.142	.131	1	21.21	.00
12	3.142	.458	1	16.39	.00
13	3.142	.785	1	11.57	.00
14	3.142	1.113	1	6.75	.00
15	3.142	1.440	1	1.93	.00
16	3.142	1.702	1	-1.93	.00
17	3.142	2.029	1	-6.75	.00
18	3.142	2.356	1	-11.57	.00
19	3.142	2.683	1	-16.39	.00
20	3.142	3.011	1	-21.21	.00
21	3.011	3.142	1	-21.61	.00
22	2.683	3.142	1	-17.79	.00
23	2.356	3.142	1	-13.98	.00
24	2.029	3.142	1	-10.16	.00
25	1.702	3.142	1	-6.34	.00
26	1.440	3.142	1	-4.49	.00
27	1.113	3.142	1	-3.70	.00
28	.785	3.142	1	-2.91	.00
29	.458	3.142	1	-2.11	.00
30	.131	3.143	1	-1.32	.00
31	.000	3.012	1	-.92	.00
32	.000	2.684	1	-.71	.00
33	.000	2.357	1	-.50	.00
34	.000	2.029	1	-.29	.00
35	.000	1.702	1	-.08	.00
36	.000	1.440	1	.08	.00
37	.000	1.113	1	.29	.00
38	.000	.785	1	.50	.00
39	.000	.458	1	.71	.00
40	.000	.131	1	.92	.00

==

*** APPROXIMATE POINT VALUES AND ERRORS -- FUNCTION FORM ***

POINT NUMBER	STATE VARIABLE	STREAM FUNCTION	ERROR
1	.1049D+01	-.4283D-01	-.4283D-01
2	.1221D+01	.4439D-01	.4439D-01
3	.1583D+01	.2734D-01	.2734D-01
4	.2213D+01	-.1657D-01	-.1657D-01
5	.3192D+01	-.3407D-01	-.3407D-01
6	.4227D+01	.2142D-01	.2142D-01
7	.5986D+01	.1090D-01	.1090D-01
8	.8560D+01	-.6198D-02	-.6198D-02
9	.1237D+02	-.7152D-02	-.7152D-02
10	.1854D+02	.2760D-02	.2760D-02
11	.2123D+02	.4625D+01	.1878D-01
12	.1638D+02	.1076D+02	-.1427D-01
13	.1155D+02	.1436D+02	-.2334D-01
14	.6753D+01	.1655D+02	.3159D-02
15	.1975D+01	.1772D+02	.4641D-01
16	-.1955D+01	.1801D+02	-.2688D-01
17	-.6792D+01	.1746D+02	-.4278D-01
18	-.1156D+02	.1612D+02	.1343D-01
19	-.1633D+02	.1401D+02	.6170D-01
20	-.2125D+02	.1107D+02	-.4090D-01
21	-.2165D+02	.7595D+01	-.3857D-01
22	-.1772D+02	.3666D+01	.7065D-01
23	-.1396D+02	.6226D+00	.1730D-01
24	-.1021D+02	-.1509D+01	-.5341D-01
25	-.6378D+01	-.2406D+01	-.4008D-01
26	-.4418D+01	-.1276D+01	.7535D-01
27	-.3689D+01	-.1149D+01	.1056D-01
28	-.2951D+01	-.1463D+01	-.4528D-01
29	-.2147D+01	-.1845D+01	-.3594D-01
30	-.1250D+01	-.2103D+01	.6794D-01
31	-.9230D+00	-.1793D+01	-.6302D-02
32	-.7570D+00	-.1170D+01	-.4865D-01
33	-.5017D+00	-.5862D+00	-.1743D-02
34	-.2471D+00	-.7394D-01	.4458D-01
35	-.6163D-01	.3230D+00	.2170D-01
36	.1845D-01	.3966D+00	-.6488D-01
37	.2927D+00	.3766D+00	.1067D-02
38	.5445D+00	.3071D+00	.4451D-01
39	.7331D+00	.1706D+00	.2478D-01
40	..8778D+00	-.4606D-01	-.3889D-01

*** APPROXIMATE NODAL VALUES AND ERRORS ***

NODE NUMBER	STATE VARIABLE	STREAM FUNCTION	ERROR
1	.9910D+00	-.1845D+00	-.8983D-02
2	.3701D+01	-.3399D-02	-.3399D-02
3	.2316D+02	-.1300D-01	.1459D-01
4	.3718D-01	.1797D+02	.3718D-01
5	-.2341D+02	.9531D+01	-.2689D+00
6	-.4762D+01	-.1938D+01	.4851D-01
7	-.8568D+00	-.2050D+01	.1432D+00
8	-.5336D-01	.4182D+00	-.5336D-01

*** BESSEL INEQUALITY ***
.2978D+04 >= .2978D+04 AND THE DIFFERENCE IS .5631D-01

7.7. Modeling Groundwater Contaminant Transport

The CVBEM may be used to develop a numerical model of contaminant transport of a conservative species in a saturated, confined groundwater aquifer. An application is to consider steady-state, two-dimensional, advection transport flow. Applications include background flows, sources and sinks, and flows introduced by boundary conditions. The numerical model produces locations of streamlines and the contaminant front locations as it changes in time. Because the CVBEM exactly solves the governing mathematical PDE, there is only error in matching prescribed boundary conditions. Because potential flow theory may be used to depict streamlines of groundwater flow for analyzing the extent of subsurface contaminant movement, potential flow theory may be used to determine if a more sophisticated study based on a long period of observation and an expensive data collection program is required. With the CVBEM, potential flow theory is used to solve analytically the groundwater flow field as governed by sources and sinks (pumping wells and recharge wells), while the background flow conditions are modeled by means of a Cauchy integral optimized with respect to minimizing ℓ_2 error in matching boundary condition values. The CVBEM technique accommodates nonhomogeneity on a regional scale (i.e., homogeneous in large subdomains of the problem), and can include spatially distributed sources and sinks such as those mathematically described by Poisson's equation. For steady state, two-dimensional, homogeneous-domain problems, the CVBEM is used to develop an approximation function which combines an exact solution of the governing groundwater flow equation (Laplace equation) and approximate solutions of the boundary conditions. For unsteady flow problems, the CVBEM can be used to give approximate solution to the time advancement of groundwater contaminants by implicit finite difference time-stepping procedures analogous to domain models.

The application of the CVBEM contaminant transport model considered is restricted to steady-state flow cases in which solute transport is by advection only. When time-dependent boundary conditions are present and dispersion-diffusion effects are significant, a steady state modeling approach becomes inappropriate. A limitation of this technique is that it does not accommodate nonhomogeneity and anisotropy within the aquifer. These complexities rapidly exceed the modeling capability of the analytic function technique.

7.7.1. Application 1A

Figure 7.10 shows a completely penetrating well pumping 50 m^3/hr from a homogeneous insotropic aquifer 10 m thick. Contaminated water is being recharged at a rate of 50 m^3/hr at a second well (injection well) located 848.5 m from the pumping well. Effective porosity is 0.25, saturated hydraulic conductivity is 1 m/hr, and negligible background groundwater flow is assumed. Figure 7.10 shows the limits of groundwater contamination corresponding to elapsed times of 0.5, 2, and 4 years. The numerical model predicts a first arrival of contamination 4.3 years after beginning of the process.

7.7.2 Application 1B

Two discharge wells are added as shown in Figure 7.11 to the system of Application 1A. The contaminant front is shown for 0.5, 2, and 4 years. It takes 4.3 years for the contaminated water to reach the middle well, and about 5.6 years for the contaminated water to reach the other two wells.

7.7.3. Application 2A

Here we will consider the steady flow pattern produced by a single pumping well (50 m^3/hr) near a landfill site with an equipotential boundary (ϕ = 2m) along the coordinate y = 1000 m. As shown in Figure 7.12, it takes the contaminant front produced by the landfill 9.0 years to reach the pumping well.

7.7.4. Application 2B

When two injection wells are installed between the landfill and the pumping well, their influence on retarding the contaminant movement can be assessed. When 10 m^3/hr is injected at each well it takes more than 13 years for the contaminant front to reach the pumping well (See Figure 7.13).

215

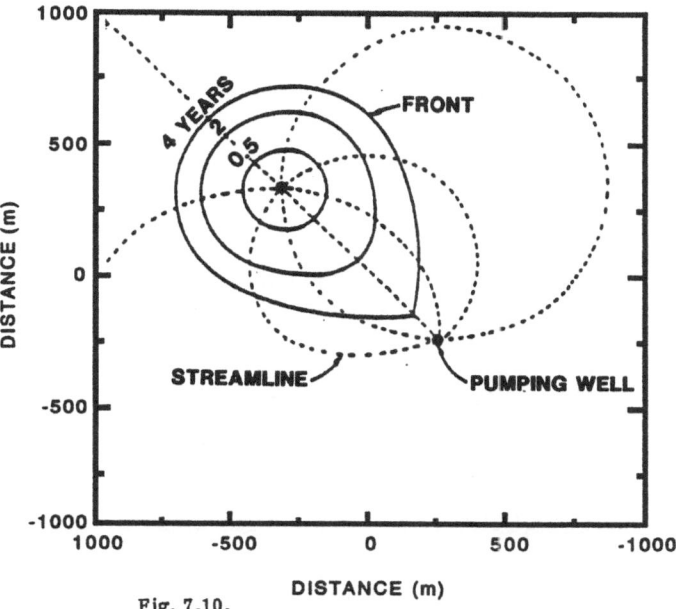

Fig. 7.10.

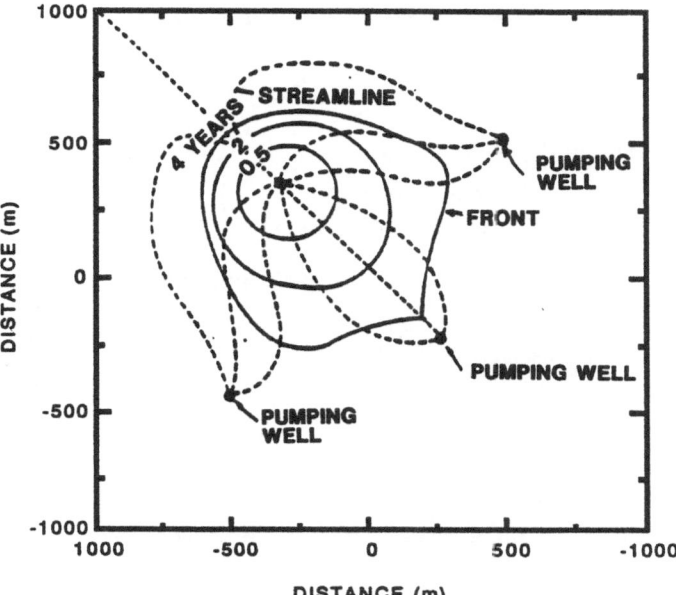

Fig. 7.11.

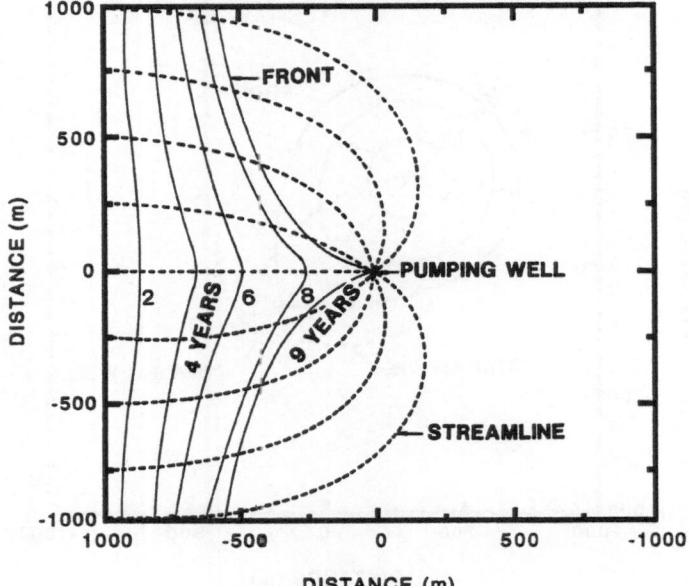

Fig. 7.12.

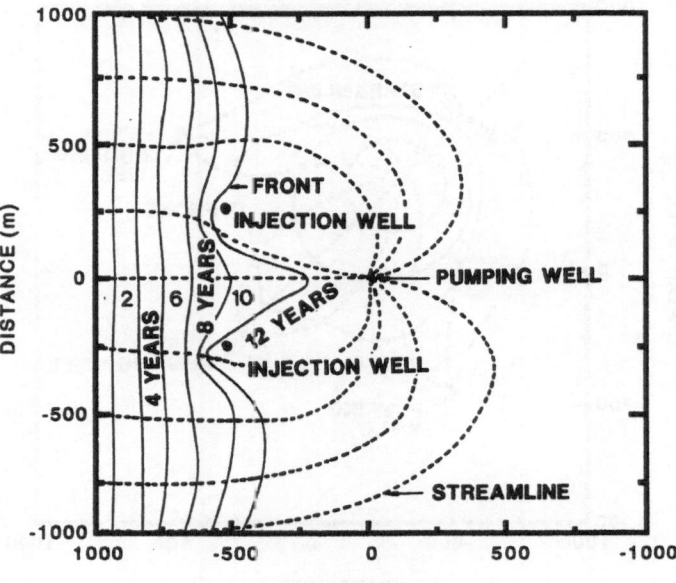

Fig. 7.13.

7.8. Three Dimensional Potential Problems

In this section, the Best Approximation Method is applied to find the best approximate solution $\hat{\phi}_m \in S_m$ to the three-dimensional Laplace equation with boundary conditions:

$$L\phi = 0 \text{ on domain } \Omega;$$

$$(7.8.1)$$

$$\phi = \phi_b \text{ on boundary } \Gamma,$$

where S_m is the m-dimensional space spanned by the m real-variable trial functions $\{f_i\}$.

By judicious choice of trial functions $\{f_i\}$, the inner-product can be simplified. In particular, by choosing trial functions $\{f_i\}$ which satisfy the linear operator equation

$$Lf_i = 0 \text{ on } \Omega \text{ for } i = 1,...,m \qquad (7.8.2)$$

the inner-product in (5.2.2) reduces to

$$(f_i, f_j) = \int_{\Gamma} f_i f_j \, d\Gamma \qquad (7.8.3)$$

Since only the problem's boundary is used in (7.8.3), the inner-product weighting, ϵ, is unnecessary.

7.8.1. Approximation Error Evaluation - Approximate Boundary Method

If $\hat{\phi}_m$ satisfies the auxiliary conditions in (7.8.1) exactly, and the trial functions are chosen that satisfy (7.8.2), then $\hat{\phi}_m = \phi$ everywhere on $\Omega \cup \Gamma$. If $\hat{\phi}_m$ is not exact, the least-squares (ℓ_2) error occurs on Γ. The Approximate Boundary Method requires that a new boundary, $\hat{\Gamma}$, be constructed which has the property that (for $\hat{\Omega}$ being the domain enclosed by $\hat{\Gamma}$),

$$L\hat{\phi}_m = 0 \text{ on } \hat{\Omega};$$

$$(7.8.4)$$

$$\hat{\phi}_m = \phi_b \text{ on } \hat{\Gamma}.$$

In this way $\hat{\phi}_m$ forms an exact solution to the problem, but with a geometrically transformed domain and boundary. An evaluation of the modeling error can be made by comparing Γ with $\hat{\Gamma}$.

7.8.2. Computer Implementation

A FORTRAN computer program was prepared for an important class of potential flow: three-dimensional steady-state heat transport. In order to develop the approximation, $\hat{\phi}_m$, each trial function, f_i is itself approximated as a finite-dimensional vector, $\mathbf{F}_i$, whose elements are the value of the trial function evaluated at each of a set of evaluation points defined along the problem boundary. Because three-dimensional steady-state heat transport is considered, the trial functions were selected for this application that satisfy the Laplacian equation $\nabla^2(f_i) = 0$ over the problem domain, Ω, thereby making ℓ^2 error minimization necessary only on the boundary, Γ. Because the trial functions $\{f_i\}$ are represented by discrete vectors, $\{\mathbf{F}_i\}$, all Gram-Schmidt operations between functions are computed by equivalent operations between vectors in $\mathbf{R}^n$. That is, the inner-product of (7.8.3) is replaced by the familiar vector dot product of two vectors:

$$(\mathbf{F}_i, \mathbf{F}_j) = \sum_{k=1}^{n} \mathbf{F}_{ik} \mathbf{F}_{jk}, \qquad (7.8.5)$$

where $\mathbf{F}_{ik}$ is the kth component of the ith vector.

The problem boundary Γ is represented as a mesh of n evaluation points stored in a single vector, $\mathbf{C}$. The vectors $\mathbf{F}_i$ are developed by evaluating trial function f_i at each (x,y,z) nodal point stored in vector $\mathbf{C}$.

Upon development of m vectors $\mathbf{F}_i$ (for m trial functions f_i), the computer program then (1) orthonormalizes the trial functions (basis vectors), and (2) computes the coefficients of the "best approximation", $\hat{\phi}_m$. The vectors, $\mathbf{F}_i$, representing the trial functions, are orthonormalized by the familiar Gram-Schmidt process. The resulting orthonormalized vectors are denoted by vectors $\mathbf{G}_i$. Note that the inner-product used is the vector dot product of Eq. (7.8.5).

Once the m orthonormalized vectors $\mathbf{G}_i$ have been constructed, the generalized Fourier coefficients are computed. The Fourier coefficients are used in back substitution through the Gram-Schmidt process to find the coefficients of the approximation $\hat{\phi}_m$. These final coefficients are written to a file for post-processing in routines that generate the approximate boundaries.

7.8.3. Application

The problem is that of finding the steady-state temperature ϕ at any point inside the shielding of a nuclear reactor. The reactor is modeled as a solid block one unit long, half a unit wide, and three-quarters of a unit tall with a spherical, high temperature, core placed nonsymmetrically in one corner (Fig. 7.14). The domain, Ω, then consists of the solid block minus the inside of the spherical core. The boundary is comprised of two disjoint parts: the surface of the block, Γ_b, and the surface of the core, Γ_c. The temperature of the outside surface of the reactor is defined to be $\phi_b = 100$ while the temperature of the inside sphere surface is set at a constant $\phi_c = 1000$. The differential equation governing the steady-state temperature distribution in the domain is the Laplacian:

$$\nabla^2\phi = \frac{\partial^2\phi}{\partial x^2} + \frac{\partial^2\phi}{\partial y^2} + \frac{\partial^2\phi}{\partial z^2} = 0 \qquad (7.8.6)$$

where $\phi = \phi(x,y,z)$ is the absolute temperature at location (x,y,z). The problem statement is thus

$$\nabla^2(\phi) = 0 \text{ in } \Omega;$$
$$\phi = 100 \text{ on } \Gamma_b; \qquad (7.8.7)$$
$$\phi = 1000 \text{ on } \Gamma_c.$$

One corner of the box is located at $(0,0,0)$ and the opposite corner is $(1.0, 0.5, 0.75)$. The center of the sphere is located at $(0.74, 0.25, 0.25)$, and its radius is 0.125 units.

7.8.4. Trial Functions

The trial functions $\{f_i\}$ chosen satisfy the Laplacian everywhere on the domain and boundary, $\Omega \cup \Gamma$. The trial functions considered are either harmonic polynomials, or the functions in (7.8.8) with a singularity, not in the domain, which is like the singularity of a Green's function

$$f(x,y,z) = \left(\frac{1}{(x-\hat{x})^2 + (y-\hat{y})^2 + (z-\hat{z})^2} \right) \qquad (7.8.8)$$

where $(\hat{x},\hat{y},\hat{z})$ is a point exterior to Ω and Γ.

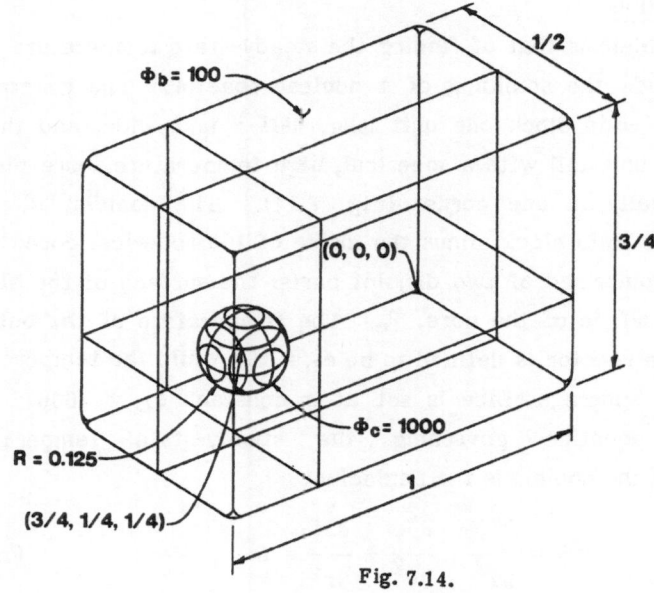

Fig. 7.14.

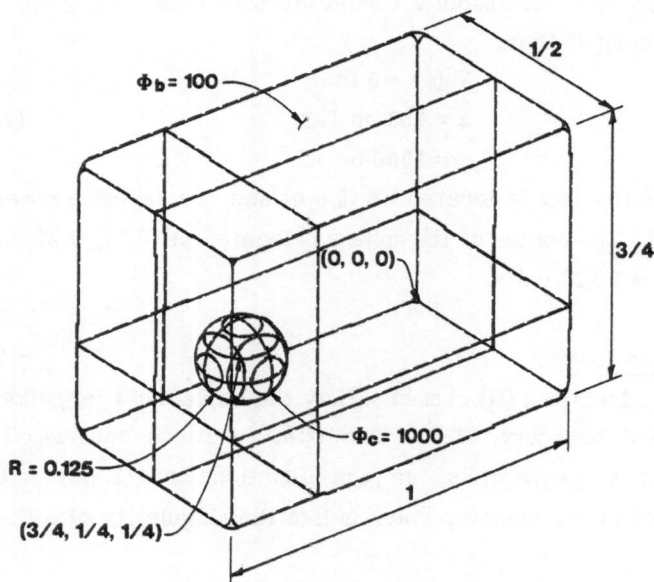

Fig. 7.15.

The set of harmonic polynomial trial functions are readily extended by scalar multiplication and addition of other harmonic polynomials. The singular functions are dependent upon the choice of singularity point location, that surround the problem outer boundary or that lie interior to the interior sphere. The sum of the several trial functions, each multiplied by a constant (to be determined), is harmonic throughout the problem domain.

Approximation accuracy is evaluated by plotting the approximate boundary $\hat{\Gamma}$ with respect to the true boundary Γ. In regions where $\hat{\Gamma}$ and Γ differ significantly, additional singular functions are added to the trial function set with singularities near that place, and harmonic polynomials are added as well.

7.8.5. Constructing the Approximate Boundary, $\hat{\Gamma}$

For each trial approximation, the approximate boundary $\hat{\Gamma}$ is developed as a set of two-dimensional slices of $\Omega \cup \Gamma$. A slice of key interest is the horizontal slice located at vertical coordinate $z = 0.25$, which corresponds to the equator or the spherical core.

For a selected set of trial functions, the model $\hat{\phi}_m$ is developed using the previous generalized Fourier series construction. The $\hat{\phi}_m(x,y,z)$ is used to evaluate the location of the isotherm $\hat{\phi}_m(x,y,z = 0.25) = 1000$ and also the location of the isotherm $\hat{\phi}_m(x,y,z = 0.25) = 100$; the resulting $\hat{\phi}_m$ isotherms are associated to boundary contours Γ_b and Γ_c, respectively.

As additional trial functions are added to the set, $\hat{\Gamma}$ approached Γ geometrically. Figure 7.15 shows the approximate boundaries $\hat{\Gamma}$ with respect to Γ. From Figure 7.15, $\hat{\Gamma}$ may be sufficiently close to Γ to suggest that the $\hat{\phi}_m$ function is adequate as an approximation of the boundary value problem. And if $\hat{\Gamma}$ is acceptable as being the "true" problem geometric shape, then $\hat{\phi}_m$ is the exact solution to the "new" boundary value problem.

CHAPTER 8
APPLICATIONS TO LINEAR OPERATOR EQUATIONS

8.0. Introduction

The Best Approximation Method has been examined, in the previous chapters, with respect to several families of linear operator equations. Chapter 7 is dedicated to an important family of linear operator equations involving potential problems (Laplace Equation). In this chapter, several families of linear operator problems will be considered and reviewed. In this way, the generalized Fourier series approach can be examined in generality.

8.1. Data Fit Analysis

In many arenas of analysis, including statistical regression, emperical equations of the form $y = f(x)$ are hypothesized to describe the relationship between dependent and independent variables, (y,x).

Let $f(x) = \lambda_1 f_1 + \lambda_2 f_2 + \cdots + \lambda_n f_n$ be the relationship prescribed between sets of values x and y where the f_i are linearly independent functions of x. Let $S = \{(x_i,y_i), i=1,2,\cdots,m\}$ be the data ordered pairs such that $m >> n$ (otherwise, if $m = n$ we have only interpolation fitting). In the set S, the ordered pairs are arranged into increasing values of the x_i such that $x_{i+1} > x_i$.

A coordinate vector C is defined by
$$C = (x_1, x_2, \cdots, x_m)$$
and an output vector is defined by
$$\Phi = (y_1, y_2, \cdots, y_m)$$
where y_j is the data value associated to x-coordinate x_j for all $j=1,2,\cdots m$. Vectors F_i are generated for each function f_i by

$$F_i = \left\{ \begin{array}{c} f_i(x_1) \\ f_i(x_2) \\ \cdot \\ \cdot \\ \cdot \\ f_i(x_{m-1}) \\ f_i(x_m) \end{array} \right\} ; \qquad i=1,2,\cdots,n$$

Then the problem of data fit becomes finding the optimized values of $\{\lambda_1, \lambda_2, \cdots, \lambda_n\}$ that minimize the ℓ_2 norm of

$$|| \Phi - \hat{\Phi}_n ||$$

where

$$\hat{\Phi}_n = \lambda_1 F_1 + \lambda_2 F_2 + \cdots + \lambda_n F_n$$

Using the dot product for the inner product, the $\{F_i\}$ are orthonormalized into elements $\{G_i\}$ by the Gram-Schmidt process, and the best approximation is

$$\hat{\Phi}_n = \lambda_1{}^* G_1 + \lambda_2{}^* G_2 + \cdots + \lambda_n{}^* G_n$$

where $\lambda_i{}^* = G_i \cdot \Phi$.

Note that this type of problem is analogous to solving over-determined matrix systems, $Ax = b$, where the column vectors in A are the vector representations F_i of basis f_i, and b is the data y_i associated to the x_i data.

8.2. Ordinary Differential Equations

DEFINITION

A **linear ordinary differential equation** or **ODE** of order n, on an interval I, has the form

$$a_0(x)y^{(n)} + a_1(x)y^{(n-1)} + \cdots + a_n(x)y^{(0)} = f(x) \qquad (8.2.1)$$

where $a_0(x), \cdots, a_n(x)$, $f(x)$ are functions of x and $y^{(0)}$ is understood to mean just y.

In operator form,

$$Ly = f(x) \qquad (8.2.2)$$

where

$$L = a_0(x) D^n + a_1(x) D^{n-1} + \cdots + a_n(x) D^0 \qquad (8.2.3)$$

where $D^n y = d^n y/dx^n$, and $D^0 y = y$.

Let $\{f_i\}$ be a set of n linearly independent functions of x on the real interval I. Then the set $\{f_i\}$ spans an n-dimensional linear space S_n. The approximation effort is to find element $\hat{\Phi}_n \in S_n$ such that $||\hat{\Phi}_n - \phi||_2$ is minimum, where ϕ solves (8.2.1) along with n auxilliary conditions.

It is recalled that the basis elements $\{f_i\}$ are linearly independent on interval I if the Wronskian $W(x) \neq 0$ for each x in the interval I where

$$W(x) = \begin{vmatrix} f_1 & f_2 & f_3 & \cdots & f_n \\ f_1' & f_2' & f_3' & \cdots & f_n' \\ f_1'' & f_2'' & f_3'' & \cdots & f_n'' \\ \cdot & & & & \cdot \\ \cdot & & & & \cdot \\ \cdot & & & & \cdot \\ f_1^{(n-1)} & f_2^{(n-1)} & f_3^{(n-1)} & \cdots & f_n^{(n-1)} \end{vmatrix} \qquad (8.2.4)$$

The inner product used is

$$(f_1, f_2) = \int_I Lf_1 Lf_2 dx + \sum_{i=1}^{n} A_i f_1 A_i f_2 \qquad (8.2.5)$$

where $A_i f_1$ is ODE auxilliary condition i applied to f_1, and n auxilliary conditions are associated with the order n ODE.

Example 8.2.1

Consider the ODE $Ly = 1 + x + x^2 + x^3 + 2x^4$ where $L = x^3 D^1 + D^0$, and $x \in [0,1]$; $y(0) = 1, y^{(1)}(0) = 1$. Let $\{f_i\} = \{1, x, x^2\}$. Evaluation points at $x = 0, 1/2, 1$ are used; that is,

$$C = (0, 1/2, 1).$$

Vectors $\{F_i\}$ are generated by three data entries as follows:

$$F_i = (A_1 f_i, A_2 f_i, Lf_i \text{ at } x = 1)$$

where, for this example $A_1 f_i = f_i(x = 0)$, $A_2 f_i = f_i'(x = 0)$.
Thus

$$f_1 = 1 \rightarrow F_1 = (1, 0, 1)$$
$$f_2 = x \rightarrow F_2 = (0, 1, 2)$$
$$f_3 = x^2 \rightarrow F_3 = (0, 0, 3)$$

The $\{F_i\}$ are linearly independent as

$$\det [F_1, F_2, F_3] = \det \begin{bmatrix} 1 & 0 & 0 \\ 0 & 1 & 0 \\ 1 & 2 & 3 \end{bmatrix} = 3 \neq 0.$$

The F_i are orthonormalized as follows:

$$\hat{G}_1 = F_1$$

$$\hat{G}_1 \cdot \hat{G}_1 = 2; \quad ||\hat{G}_1|| = \sqrt{2}$$

$$G_1 = (1, 0, 1)/\sqrt{2}$$

$$(F_2, G_1) = \sqrt{2}$$

$$\hat{G}_2 = F_2 - (F_2, G_1)G_1 = (-1, 1, 1)$$

$$(\hat{G}_2, \hat{G}_2) = 3; \quad ||\hat{G}_2|| = \sqrt{3}$$

$$G_2 = (-1, 1, 1)/\sqrt{3}$$

$$(F_3, G_1) = 3/\sqrt{2}$$

$$(F_3, G_2) = 3/\sqrt{3}$$

$$\hat{G}_3 = F_3 - (F_3, G_1)\,G_1 - (F_3, G_2)\,G_2$$

$$= (-1/2, -1, 1/2)$$

$$(\hat{G}_3, \hat{G}_3) = 3/2; \quad ||\hat{G}_3|| = \sqrt{3/2}$$

$$G_3 = (-1/2, -1, 1/2)/\sqrt{3/2}$$

$$\Phi = (A_1 y, A_2 y, Ly \text{ at } x = 1)$$

$$= (1, 1, 6)$$

The generalized Fourier series coefficients to the $\{G_i\}$ are

$$\lambda_1{}^* = (G_1, \Phi) = 7/\sqrt{2}$$

$$\lambda_2{}^* = (G_2, \Phi) = 6/\sqrt{3}$$

$$\lambda_3{}^* = (G_3, \Phi) = \sqrt{3/2}$$

From the above derivation,

$$\lambda_3{}^* G_3 = \hat{G}_3 = F_3 - (3/\sqrt{2})G_1 - (3/\sqrt{3})G_2$$

$$(-3/\sqrt{3} + \lambda_2{}^*)G_2 = \hat{G}_2 = (F_2 - \sqrt{2}\,G_1)$$

$$(-3/\sqrt{3} - \sqrt{2} + \lambda_1{}^*)G_1 = (\sqrt{2})G_1$$

Recall $G_1 = F_1/\sqrt{2}$. Adding, $\lambda_3{}^* G_3 + \lambda_2{}^* G_2 + \lambda_1{}^* G_1 = F_3 + F_2 + F_1$, giving that $(\lambda_1, \lambda_2, \lambda_3) = (1, 1, 1)$ or that with respect to the original basis $\{f_i\}$, the best approximation (as estimated using vector representations in $\mathbf{R}^3$) is

$$\hat{\phi}_3 = 1 + x + x^2$$

which is the exact solution. Note that such a modest computational effort in ℓ_2 as using vectors in $\mathbf{R}^3$ is only possible due to, among other factors, $y \in S_3$ where S_3 is spanned by the $\{f_i\}$, and due to the $\{F_i\}$ being linearly independent.

Notwithstanding, highly complex linear ODEs can be attacked using small dimension $\mathbf{R}^k$ vector representations if the $\{F_i\}$ are linearly independent, and the exact solution, y, is in the linear space spanned by the original basis functions $\{f_i\}$.

8.3. Best Approximations of Functions

Let f be defined on a region Ω (such as an interval). Let $\{f_i\}$ be a basis that spans the linear space S_n. The goal is to find the best approximation $\hat{\phi}_n \in S_n$ of f. The linear operator for this type of problem is the identity differential operator, D^0,

$$L\phi = D^0\phi \tag{8.3.1}$$

and the inner product to use is

$$(u,v) = \int_\Omega Lu\ Lv\ d\Omega = \int_\Omega u\ v\ d\Omega \tag{8.3.2}$$

In this family of linear operator equations, a large body of literature has been developed for the case of Ω being a real interval; namely, Fourier series. It is recalled that Fourier series typically require an infinite dimensional basis $\{f_i\}$ in order for the best approximation to converge to the target function. It is also recalled that even though a best approximation $\hat{\phi}_n$ for ϕ is obtained, care must be taken in using $\hat{\phi}_n$ as a substitute for ϕ such as demonstrated in Example 3.2.7.

Example 8.3.1

In general, a Taylor series of f(x) is developed about expansion point x = 0 by

$$f(x) = f(0) + f'(0)x + f''(0)\ x^2/2!+\ldots$$

where f"(0) is notation for evaluating f"(x) at x = 0. For f(x) = sinx,

$$f(x) = x - x^3/3! + x^5/5!+\ldots$$

or for a truncated cubic polynomial, $f(x) \approx x - x^3/6$. Using $\{f_i\} = \{x,\ x^3\}$ as a basis for S_2, the best approximation may be examined on an interval, say $\Omega = [0,\ \pi/2]$. The $\{f_i\}$ are orthonormalized into the new basis $\{g_i\}$ using inner product (8.3.2). The best approximation in S_2 on Ω is $\phi_2 = 0.9988x - 0.1451x^3$.

Comparing the best approximation on Ω to the Taylor series, it is seen that the best approximation $\hat{\phi}_2$ provides overall better accuracy on $[0, \pi/2]$.

<hr>

Table 8.1. Taylor Series vs. Best Approximation for sinx

x(radians)	sinx	$(x - x^3/6)$ Taylor Series	$(0.9988x-0.1451x^3)$ Best Approximation
0	0.0	0.0	0.0
0.25	0.247	0.247	0.247
0.50	0.480	0.479	0.481
0.75	0.681	0.680	0.687
1.0	0.841	0.833	0.854
1.25	0.949	0.924	0.965
1.50	0.998	0.938	1.000
$\pi/2$	1.000	0.924	1.000

<hr>

Example 8.3.2

The previous example can be analyzed with use of vectors $\{F_1, F_2\}$ in $\mathbf{R}^3$ instead of $\{f_1, f_2\}$. Let the evaluation points in Ω be

$$C = (0, 0.5, 1.5)$$

Then

$f_1 = x \rightarrow F_1 = (0, 0.5, 1.5)$

$f_2 = x^3 \rightarrow F_2 = (0, 0.125, 3.375)$

For sinx,

$\sin x \rightarrow \Phi = (0, 0.480, 0.998)$

Orthonormalizing the $\{F_i\}$ using the dot product,

$\hat{G}_1 = F_1$

$(\hat{G}_1, \hat{G}_1) = 2.5; \ ||\hat{G}_1|| = 1.5811$

$G_1 = (0, 0.3162, 0.9407)$

$(G_1, F_2) = 3.2144$

$\hat{G}_2 = (0, 0.125, 3.375) - (3.2144)(0, 0.3162, 0.9407)$

$= (0, -0.8914, 0.3514)$

$(\hat{G}_2, \hat{G}_2) = 0.9181; \ ||\hat{G}_2|| = 0.9582$

$G_2 = (0, -0.9303, 0.3667)$

For the generalized Fourier coefficients,

$$\lambda_1{}^* = (G_1, \Phi) = 1.0906$$
$$\lambda_2{}^* = (G_2, \Phi) = -0.0806$$

The best approximation $\hat{\phi}_2$ is

$$\hat{\phi}_2 = (1.0906)\, G_1 + (-0.0806)\, G_2$$
$$= (-0.0841)\, \mathbf{F}_2 + (0.8608)\, \mathbf{F}_1$$

Or,

$$\hat{\phi}_2 \cong 0.8608x - 0.0841x^3$$

which is a loss in approximation accuracy over the previous problem results.

8.4. Matrix Systems

Matrix systems $\mathbf{Ax} = \mathbf{b}$ are a familiar linear operator equation where $\mathbf{x} \in \mathbf{R}^n$, $L = \mathbf{A}$, and $\mathbf{A}$ is a mxn matrix, and $\mathbf{b} \in \mathbf{R}^m$. Chapter 1 sections 3 and 4 examine matrix system properties in detail, while Chapter 3, section 2 considers ℓ_2 optimization in finding best approximations.

The linear operator is, for $\mathbf{x} \in \mathbf{R}^n$,

$$L\mathbf{x} = \mathbf{Ax} \tag{8.4.1}$$

where the inner product used is the usual dot product.

It is noted that $||L\mathbf{x} - \mathbf{b}||_2$ - being minimum is equivalent to minimizing

$$||L\mathbf{x} - \mathbf{b}||_2^2 = (L\mathbf{x} - \mathbf{b}, L\mathbf{x} - b)$$

The matrix $\mathbf{A}$ may be resolved into column vectors $V_i \in \mathbf{R}^m$, $i = 1, 2, \cdots, n$, so that

$$\mathbf{A} = \left[\, V_1, V_2, \cdots, V_n \,\right] \tag{8.4.2}$$

Then

$$\mathbf{Ax} = x_1 V_1 + x_2 V_2 + \cdots + x_n V_n \tag{8.4.3}$$

which is now a problem posed as the familiar linear identity operator.

That is, find the values of $\{x_1, x_2, \cdots, x_n\}$ that minimize the ℓ_2 norm

$$||\, x_1 V_1 + x_2 V_2 + \cdots + x_n V_n - b\,||_2 \qquad (8.4.4)$$

where the V_i and b are given column vectors of dimension mx1. To proceed, the vectors $\{V_i\}$ are orthonormalized into vectors $\{G_i\}$ (where the V_i are linearly independent) using the dot product as the inner product. The generalized Fourier coefficients are $\lambda_i^* = G_i \cdot b$, $i=1,2,\cdots,n$. Back-substitution generates the x_i values.

Example 8.4.1 (Hilbert Matrix)

A classic matrix system problem is the Hilbert matrix $Ax = b$ where

$$A_n = \begin{bmatrix} 1 & 1/2 & 1/3 & 1/4 & 1/5 & \cdots & \dfrac{1}{n} \\ 1/2 & 1/3 & 1/4 & \cdots & & & \cdot \\ 1/3 & 1/4 & 1/5 & \cdots & & & \cdot \\ \cdot & & & & & & \cdot \\ \cdot & & & & & & \\ \cdot & & & & & & \\ \dfrac{1}{n} & \dfrac{1}{n+1} & & \cdots & & & \dfrac{1}{2n-1} \end{bmatrix}$$

For $n = 5$, $A_5 x$ can be written as

$$A_5 x = x_1 \begin{Bmatrix} 1 \\ 1/2 \\ 1/3 \\ 1/4 \\ 1/5 \end{Bmatrix} + x_2 \begin{Bmatrix} 1/2 \\ 1/3 \\ 1/4 \\ 1/5 \\ 1/6 \end{Bmatrix} + x_3 \begin{Bmatrix} 1/3 \\ 1/4 \\ 1/5 \\ 1/6 \\ 1/7 \end{Bmatrix} + x_4 \begin{Bmatrix} 1/4 \\ 1/5 \\ 1/6 \\ 1/7 \\ 1/8 \end{Bmatrix} + x_5 \begin{Bmatrix} 1/5 \\ 1/6 \\ 1/7 \\ 1/8 \\ 1/9 \end{Bmatrix}$$

Although manageable in appearance, the Hilbert matrix system is associated with instability in that a uniform error ϵ in b can result in a 500,000-fold increase in error in the estimates of x (for the case of $n = 5$). The instability issue becomes apparent when the inverse Hilbert matrix A_n^{-1} is determined, and $||A_n^{-1}||$ is computed (see section 4.2).

8.5. Linear Partial Differential Equations

Let $z = z(x,y)$ be a function of two variables. Then a partial differential equation (PDE) of the form

$$x^2 \frac{\partial^4 z}{\partial x^4} + y^2 \frac{\partial^4 z}{\partial x^2 \partial y^2} + xy \frac{\partial^4 z}{\partial y^4} + xy^2 \frac{\partial z}{\partial x} = x^2 e^y \qquad (8.5.1)$$

is an example of a fourth order linear PDE.

The two-dimensional Laplace equation is

$$\frac{\partial^2 \phi}{\partial x^2} + \frac{\partial^2 \phi}{\partial y^2} = 0 \qquad (8.5.2)$$

In three-dimensions,

$$\frac{\partial^2 \phi}{\partial x^2} + \frac{\partial^2 \phi}{\partial y^2} + \frac{\partial^2 \phi}{\partial z^2} = 0 \qquad (8.5.3)$$

Similarly, a three-dimensional Poisson equation is

$$\frac{\partial^2 \phi}{\partial x^2} + \frac{\partial^2 \phi}{\partial y^2} + \frac{\partial^2 \phi}{\partial z^2} = f(x,y,z)$$

The above linear PDE can be modeled using the Best Approximation method.

Example 8.5.1

A steady-state transport of specie ϕ in the unit square domain $\Omega = (x,y): 0 \leq x \leq 1, 0 \leq y \leq 1$ is described by

$$\frac{\partial^2 \phi}{\partial x^2} + \frac{\partial^2 \phi}{\partial y^2} = 2$$

On the boundary, Γ, ϕ_b is measured as

$$\phi_b = \begin{cases} 1 + x; & y = 0 \\ 1 + y^2; & x = 0 \\ 2 + y^2; & x = 1 \\ 2 + x; & y = 1 \end{cases}$$

The basis functions considered are $\{f_i\} = \{1, x, y^2\}$. Representative vectors F_i are generated by use of four evaluation points on Γ to model ϕ_b, and a single evaluation point in the center of Ω to model $L\phi$ in Ω. The vector C of evaluation point coordinates is

$$C = ((0.5, 0), (1, 0.5), (0.5, 1), (0, 0.5), (0.5, 0.5))$$

The vector to be approximated is

$$\Phi = (\phi_b(0.5, 0), \phi_b(1, 0.5), \phi_b(0.5, 1), \phi_b(0, 0.5), L\phi(0.5, 0.5))$$
$$= (1.5, 2.25, 2.5, 1.25, 2)$$

The resulting vectors, $\{F_i\}$, generated are

$$f_1 = 1 \rightarrow F_1 = (1, 1, 1, 1, 0)$$
$$f_2 = x \rightarrow F_2 = (0.5, 1, 0.5, 0, 0)$$
$$f_3 = y^2 \rightarrow F_3 = (0, 0.25, 1, 0.25, 2)$$

The Gram–Schmidt process produces the following orthonormalization for $\{F_i\}$:

$$\hat{G}_1 = F_1$$
$$(\hat{G}_1, \hat{G}_1) = 4; \quad \|\hat{G}_1\| = 2$$
$$G_1 = (1, 1, 1, 1, 0)/2$$
$$(G_1, F_2) = 1$$
$$\hat{G}_2 = F_2 - (G_1, F_2)G_1 = (0, 0.5, 0, -0.5, 0)$$
$$(\hat{G}_2, \hat{G}_2) = 0.5; \quad \|\hat{G}_2\| = \sqrt{2}/2$$
$$G_2 = (0, 1/\sqrt{2}, 0, -1/\sqrt{2}, 0)$$
$$(G_1, F_3) = 3/4$$
$$(G_2, F_3) = 0$$
$$\hat{G}_3 = F_3 - (G_1, F_3)G_1 - (G_2, F_3)G_2 = (-3, -1, 5, -1, 16)/8$$
$$(\hat{G}_3, \hat{G}_3) = 73/16; \quad \|\hat{G}_3\| = \sqrt{73}/4$$
$$G_3 = (-3, -1, 5, -1, 16)/2\sqrt{73}$$

The generalized Fourier series coefficients are computed as

$$\lambda_1^* = (\Phi, G_1) = 15/4$$
$$\lambda_2^* = (\Phi, G_2) = 1/\sqrt{2}$$
$$\lambda_3^* = (\Phi, G_3) = \sqrt{73}/4$$

The best approximation is $\hat{\phi}_3 = \lambda_1^* G_1 + \lambda_2^* G_2 + \lambda_3^* G_3$. But

$$G_3 = (4/\sqrt{73}) F_3 - (3/\sqrt{73}) G_1$$
$$G_2 = (\sqrt{2}) F_2 - (\sqrt{2}) G_1$$
$$G_1 = F_1/2$$
$$\therefore \hat{\phi}_3 = F_3 + F_2 + F_1$$

implying the best approximation with respect to the original basis $\{1, x, y^2\}$ is

$$\hat{\phi}_3 = 1 + x + y^2$$

which is the exact solution.

Example 8.5.2 (Nonhomogeneous and Anisotropic Transport)

For the same domain Ω described in the previous example, suppose ϕ is governed by the PDE

$$\frac{\partial}{\partial x} K_1(x) \frac{\partial \phi}{\partial x} + \frac{\partial}{\partial y} K_2(y) \frac{\partial \phi}{\partial y} = 5$$

where $K(x)$ and $K(y)$ are specie transport conductivities defined in Ω by

$$K_1(x) = 1 + x; \; K_2(y) = 2; \; (x,y) \text{ in } \Omega.$$

The measured values of ϕ on the boundary Γ are as given in the previous example. Then the governing PDE is

$$\frac{\partial}{\partial x} (1+x) \frac{\partial \phi}{\partial x} + 2 \frac{\partial^2 \phi}{\partial y^2} = 5$$

Using the same basis $\{f_i\} = \{1, x, y^2\}$, and the dot product for the inner product, vectors $\{F_i\}$ are constructed below.

The linear operator L is

$$L = \frac{\partial}{\partial x} (1+x) \frac{\partial}{\partial x} + 2 \frac{\partial^2}{\partial y^2}$$

where $L\phi = 5$ in Ω. The vector of evaluation points is the same C vector used previously. The vector to be approximated is now

$$\Phi = (1.5, 2.25, 2.5, 1.25, 5)$$

Note that $\{Lf_i\} = \{0, 1, 4\}$, for $i = 1, 2, 3$.

Then

$$f_1 = 1 \rightarrow F_1 = (1, 1, 1, 1, 0)$$
$$f_2 = x \rightarrow F_2 = (0.5, 1, 0.5, 0, 1)$$
$$f_3 = y^2 \rightarrow F_3 = (0, 0.25, 1, 0.25, 4)$$

Orthonormalizing the $\{F_i\}$,

$$\hat{G}_1 = F_1$$

$$(\hat{G}_1, \hat{G}_1) = 4; \ ||\hat{G}_1|| = 2$$

$$G_1 = (1, 1, 1, 1, 0)/2$$

$$(F_2, G_1) = 1$$

$$\hat{G}_2 = F_2 - (F_2, G_1)G_1 = (0, ,0.5, 0, -0.5, 1)$$

$$(\hat{G}_2, \hat{G}_2) = 3/2; \ ||\hat{G}_2|| = \sqrt{3/2}$$

$$G_2 = (0, 0.5, 0, -0.5, 1)/\sqrt{3/2}$$

$$(F_3, G_1) = 3/4$$

$$(F_3, G_2) = 4/\sqrt{3/2}$$

$$\hat{G}_3 = F_3 - (F_3, G_1)G_1 - (F_3, G_2)G_2$$

$$= (-3/8, -35/24, 5/8, 29/24, 4/3)$$

$$(\hat{G}_3, \hat{G}_3) = 283/48; \ ||\hat{G}_3|| = \sqrt{283/48}$$

$$G_3 = (-9, -35, 15, 29, 32)/24\sqrt{283/48}$$

The generalized Fourier series coefficients are

$$\lambda_1^* = (G_1, \phi) = 15/4$$

$$\lambda_2^* = (G_2, \phi) = 11/\sqrt{6}$$

$$\lambda_3^* = (G_3, \phi) = \sqrt{283/48}$$

The best approixmation $\hat{\phi}_3$ is

$$\hat{\phi}_3 = \lambda_3^* G_3 + \lambda_2^* G_2 + \lambda_1^* G_1$$

$$= F_1 + F_2 + F_3$$

implying that $(\lambda_1, \lambda_2, \lambda_3) = (1, 1, 1)$, and $\hat{\phi}_3 = 1 + x + y^2$ which is the exact solution to the PDE and auxilliary conditions.

8.6. Linear Integral Equations

Many integral equations satisfy the properties of being linear operators. One important type of integral equation occurs in the modeling of flood flows, the convolution integral.

Let $e(t)$ be the effective rainfall (i.e., rainfall less any amounts absorbed or stored in the drainage area) on the drainage area Ω of area A, and let $q(t)$ be the runoff hydrograph (time distribution of flows measured in volume/time). An integral equation that links $(e(t), q(t))$, for only mild conditions imposed, (such as, $e(t)$ begins to be nonzero at a time prior or equal to when $q(t)$ becomes nonzero) is the convolution

$$q(t) = \int_{s=0}^{t} e(t-s)\ \psi(s)\ ds \tag{8.6.1}$$

where $\psi(s)$ is a transfer function that is to be determined.

The integral is a linear operator where for an element ϕ

$$L\phi = \int_{s=0}^{t} e(t-s)\ \phi(s)\ ds \tag{8.6.2}$$

8.6.1. An Inverse Problem

The integral equation can be attacked as an inverse problem. To proceed numerically, a timestep Δ is used to approximate (8.6.1) by step functions,

$$q(t) \cong \sum_{i=1}^{m} q_i\ \eta_i(t) \tag{8.6.3}$$

where $m\Delta$ is a large duration that more than encompasses the total $e(t)$ storm time duration, $\eta_i(t)$ is the characteristic function (see Example 1.2.6); and q_i is the mean value of $q(t)$ in time interval $I_i = [(i-1)\Delta, \Delta]$.

Similarly, using time increment Δ, $e(t)$ is discretized into

$$e(t) \cong \sum_{i=1}^{m} e_i\ \eta_i(t) \tag{8.6.4}$$

where e_i is the mean value of $e(t)$ in I_i.

Finally,

$$\psi(t) \cong \sum_{i=1}^{m} \psi_i\ \eta_i(t) \tag{8.6.5}$$

In matrix form, the above equations are related by the system

$$
\begin{bmatrix}
e_1 & 0 & 0 & \cdots & & & 0 \\
e_2 & e_1 & 0 & & & & \cdot \\
e_3 & e_2 & e_1 & & & & \cdot \\
\cdot & e_3 & e_2 & & & & \cdot \\
\cdot & \cdot & e_3 & & & & \cdot \\
 & & \cdot & & e_1 & & \cdot \\
 & & & & e_2 & & \\
 & & & & e_3 & & \\
e_m & e_{m-1} & e_{m-2} & & \cdot & & \\
0 & e_m & e_{m-1} & & & & \\
0 & 0 & e_m & & & & \\
0 & 0 & 0 & & & & \\
\cdot & \cdot & \cdot & & & & \\
0 & 0 & 0 & \cdots & & & e_m
\end{bmatrix}_{p \times n}
\left\{
\begin{array}{c}
\psi_1 \\ \psi_2 \\ \psi_3 \\ \psi_4 \\ \cdot \\ \cdot \\ \cdot \\ \\ \\ \\ \\ \psi_n
\end{array}
\right\}_{n \times 1}
=
\left\{
\begin{array}{c}
q_1 \\ q_2 \\ q_3 \\ q_4 \\ \cdot \\ \cdot \\ \cdot \\ \\ \\ \\ \\ q_{p-1} \\ q_p
\end{array}
\right\}_{p \times 1}
\tag{8.6.6}
$$

where $p = m+n-1$. Note that each e_i and q_i are nonzero values.

In column vector form, (8.6.6) is

$$
\begin{bmatrix} e_1, e_2, \cdots, e_1 \end{bmatrix}
\left\{
\begin{array}{c}
\psi_1 \\ \psi_2 \\ \cdot \\ \cdot \\ \cdot \\ \psi_n
\end{array}
\right\}
= q
\tag{8.6.7}
$$

where each e_i and q are $p \times 1$ column vectors.

Then the best approximation $\hat{\phi}_n$ is

$$\hat{\phi}_n = \psi_1\, e_i + \psi_2\, e_2 + \cdots + \psi_n\, e_n$$

and the goal is to minimize the ℓ_2 norm

$$||\hat{\phi}_n - q||_2 \tag{8.6.8}$$

From (8.6.7), the vectors e_i form a basis of linear space V, and the best approximation $\hat{\phi}_n \in V$ is sought that minimizes the norm of (8.6.8). The ψ_i values are computed from the methods of section 8.4. The resulting ψ_i values, $i=1,2,\cdots,n$ are then used to define the step function representation of $\psi(s)$.

It is noted that when using step functions, the vector representations $\{F_i\}$ and Φ are readily determined.

8.6.2. Best Approximation of The Transfer Function in a Linear Space

Given a linear space S_k spanned by k linearly independent functions forming a basis for S_k, given by $\{f_i\}$, then an element $\hat{\phi}_k \in S_k$ is

$$\hat{\phi}_k = c_1 f_1 + c_2 f_2 + \cdots + c_k f_k$$

Also, $L\hat{\phi}_k = \sum_{i=1}^{k} c_i L f_i$. Generally, Lf_i is numerically estimated such as by use of the previous matrix system of (8.6.6). That is, Lf_i is the convolution of $e(t)$ and $f_i(t)$,

$$Lf_i(t) = \int_{s=0}^{t} e(t-s)\, f_i(s)\, ds \tag{8.6.9}$$

Using time increment Δ,

$$f_i(t) \cong \sum_{1}^{m} f_i\, X_i(t) \tag{8.6.10}$$

and $Lf_i(t)$ is the vector, F_i, given by

$$F_i = [e_1, \ e_2, \ \cdots, \ e_n] \begin{Bmatrix} f_1 \\ f_2 \\ . \\ . \\ . \\ f_n \end{Bmatrix} \qquad (8.6.11)$$

where f_i is the mean value of $f(t)$ in the time interval I_i. (Note that a common dimension n is used with several of the vectors above in order to utilize matrices.) Then the best approximation vector $\hat{\phi}_k = \lambda_1 F_1 + \lambda_2 F_2 + \cdots + \lambda_k F_k$ is found by minimizing the ℓ_2 norm, $||\hat{\phi}_k - \phi||$. Given the $\{\lambda_i\}$, the f_i basis is used to get

$$\hat{\phi}_k = \lambda_1 f_1 + \lambda_2 f_2 + \cdots + \lambda_k f_k \qquad (8.6.12)$$

When using vectors, $\{F_i\}$, the dot product is used as the inner product to orthonormalize the $\{F_i\}$. When using the original basis elements, $\{f_i\}$, the L_2 inner product is used to orthonormalize the basis elements,

$$(v,w) = \int_{t=0}^{\infty} Lv \ Lw \ dt \qquad (8.6.13)$$

$$= \int_{t=0}^{\infty} \left(\int_{s_1=0}^{t} e(t-s_1) \ v(s_1) ds_1 \right) \left(\int_{s_2=0}^{t} e(t-s_2) \ w(s_2) ds_2 \right) dt$$

where the product $LuLv$ generally involves numerical integration to evaluate. A comparison of (8.6.13) to (8.6.11) depicts the approximation of the above Lv and Lw terms by use of matrices. The dot product between vectors F_v and F_w approximates the temporal integral of (8.6.13) over time domain $(0, t_\infty)$. Example 5.3.2 provides an illustration of approximating the transfer function, $\psi(s)$, as by use of a best approximation in a linear space.

REFERENCES

1. Birkhof, G. and Lynch, R., Numerical Solution of Elliptic Problems, SIAM Studies in Applied Math, 1984.

2. Hromadka II, T.V., Pinder, G., and Joos, B., Approximating a Linear Operator Equation Using a Generalized Fouries Series: Development, Engineering Analysis, 1987 4(1).

3. Davis, P.J. and Rabinowitz, P., Advances in Orthonormalizing Computation, Academic Press, 1961.

4. Duren, Peter L., Theory of H^p Spaces, Academic Press, New York, 1970.

5. Frink, E.O., Matanga, G.B. and Cherry, J.A., The Dual Formulation of Flood for Contaminant Transport Modeling, 2, The Borden Aquifer, Water Resources Research, Vol. 21, No. 2, 1985.

6. Hromadka II, T.V., the Complex Variable Boundary Element Method, Springer-Verlag, New York, 1984.

7. Hromadka II, T.V., Lai, Chintu and Yen, C.C., A Complex Boundary Element Model of Flow-Field Problems Without Matrices, submitted to Engineering Analysis, 1986.

8. Hromadka II, T.V., Pinder, G.F. and Joos, B., Approximating a Linear Operator Equation Using a Generalized Fourier Series: Development, in-review, 1986.

9. Hromadka II, T.V. and Yen, C.C., Complex Boundary Element Model of Flow Field Problems Without Matrices, Engineering Analysis, 1987 4(1).

10. Kantorovich, L.V. and Krylov, V.I., Approximate Methods of Higher Analysis, Interscience Publishers, New York, 1964.

11. Lapidus, Leon, and Pinder, G.F., Numerical Solution of Partial Differential Equation in Science and Engineering, John Wiley & Sons, 1982.

12. Mathews, J.H., Basic Complex Variables for Mathematics and Engineering, Allyn and Bacon, Inc., Boston, Mass., 1982.

APPENDIX A
DERIVATION OF CVBEM APPROXIMATION FUNCTION

Let $\omega \in W_\Omega A$, and P_n be a nodal partition of Γ. Define a global trial function $G_m(\zeta)$ on Γ by

$$G_m(\zeta) = \sum_{j=1}^{m} N_j(\zeta)\, \omega_j$$

where $\omega_j = \omega(z_j)$ and $\zeta \in \Gamma$. Develop the integral function $A(z)$ defined by

$$A(z) = \frac{1}{2\pi i} \int_{\Gamma} \frac{G_m(\zeta)\, d\zeta}{\zeta - z}, \qquad z \in \Omega$$

$$= \sum_{j=1}^{m} \frac{1}{2\pi i} \int_{\Gamma_j} \frac{G_m(\zeta)\, d\zeta}{\zeta - z}$$

On Γ_j, $G_m(\zeta) = \omega_j \left[(z_{j+1} - \zeta)/(z_{j+1} - z_j) \right]$

$$+ \omega_{j+1} \left[(\zeta - z_j)/(z_{j+1} - z_j) \right]$$

and

$$\int_{\Gamma_j} \frac{G_m(\zeta)\, d\zeta}{\zeta - z} = \omega_{j+1} \left[1 + \left(\frac{z - z_j}{z_{j+1} - z_j} \right) (\operatorname{Ln}(z_{j+1} - z) - \operatorname{Ln}(z_j - z)) \right]$$

$$- \omega_j \left[1 + \left(\frac{z - z_{j+1}}{z_{j+1} - z_j} \right) (\operatorname{Ln}(z_{j+1} - z) - \operatorname{Ln}(z_j - z)) \right]$$

Summing from $j = 1$ to m, (and noting $\omega_{m+1} = \omega_1$ and $z_{m+1} = z_1$)

$$\chi = \sum_{j=1}^{m} \int_{\Gamma_j} \frac{G_m(\zeta)d\zeta}{\zeta - z} = \sum_{j=1}^{m} (\omega_{j+1} - \omega_j)$$

$$+ \sum_{j=1}^{m} \frac{\left[\{\omega_{j+1}(z - z_j) - \omega_j(z - z_{j+1})\}(Ln\ (z_{j+1} - z) - Ln(z_j - z))\right]}{(z_{j+1} - z_j)}$$

where $Ln(z_{m+1} - z) = Ln(z_1 - z) + 2\pi i$.

Thus

$$\chi = \sum_{j=1}^{m-1} \left[\omega_{j+1}(z - z_j) - \omega_j(z - z_{j+1})\right]\left[Ln(z_{j+1} - z) - Ln(z_j - z)\right]/(z_{j+1} - z_j)$$

$$+ \left[\omega_1(z - z_m) - \omega_m(z - z_1)\right]\left[Ln(z_1 - z) + 2\pi i - Ln(z_m - z)\right]/(z_1 - z_m)$$

Combining terms with respect to the $Ln(z_j - z)$ functions gives

$$\chi = \sum_{j=1}^{m} \left(\frac{[\omega_j(z - z_{j-1}) - \omega_{j-1}(z - z_j)]}{(z_j - z_{j-1})} - \frac{[\omega_{j+1}(z - z_j) - \omega_j(z - z_{j+1})]}{(z_{j+1} - z_j)} \right) Ln(z_j - z)$$

$$+ 2\pi i \frac{[\omega_1(z - z_m) - \omega_m(z - z_1)]}{(z_1 - z_m)}$$

Thus if $P_j(\zeta)$ is the interpolation function on Γ_j given by

$$P_j(\zeta) = \begin{cases} N_j(\zeta)\omega_j + N_{j+1}(\zeta)\omega_{j+1}, & \zeta \in \Gamma_j \\ 0, & \text{otherwise} \end{cases}$$

then by substituting z into ζ of $P_j(\zeta)$

$$X = \sum_{j=1}^{m} (P_{j-1}(z) - P_j(z)) Ln(z_j - z) + 2\pi i\, P_m(z)$$

where now $P_0(z) \equiv P_m(z)$. Finally, $P_j(z_j) = P_{j-1}(z_j)$ implies that

$$A(z) = P_m(z) + \frac{1}{2\pi i} \sum_{j=1}^{m} \left[\frac{(\omega_{j+1} - \omega_j)}{(z_{j+1} - z_j)} - \frac{(\omega_j - \omega_{j-1})}{(z_j - z_{j-1})} \right] (z_j - z)\, Ln\, (z_j - z)$$

Complex constants k_j can be used to simplify the writing of $A(z)$ by

$$A(z) = P_m(z) + \sum_{j=1}^{m} k_j\, (z_j - z)\, Ln\, (z_j - z)$$

where

$$k_j = \frac{1}{2\pi i} \left[\frac{(\omega_{j+1} - \omega_j)}{(z_{j+1} - z_j)} - \frac{(\omega_j - \omega_{j-1})}{(z_j - z_{j-1})} \right]$$

Noting $Ln\, (z_j - z) = Ln\, (z - z_j) + Ln\, (-1) = Ln\, (z - z_j) + i\pi$, $A(z)$ can be rewritten as

$$A(z) = P_m(z) - \sum_{j=1}^{m} k_j(z - z_j)\, Ln\, (z - z_j) + i\pi \sum_{j=1}^{m} k_j(z_j - z)$$

In the above, $Ln\, (z - z_j)$ is measured with respect to point $z \in \Omega$ as the branch point. It is desirable to define branch points at each z_j with branch cuts lying exterior of Ω. This process introduces an additional angle term θ^j of the branch cut for each node such that

$$Ln\, (z - z_j) = Ln_j\, (z - z_j) + i\theta^j$$

where Ln_j is notation of individual logarithm functions.
Thus $A(z)$ is of the form

$$A(z) = R_1(z) + \sum_{j=1}^{m} c_j(z - z_j)\, Ln_j(z - z_j)$$

where $R_1(z)$ is a first degree complex polynomial

$$R_1(z) = P_m(z) + \sum_{j=1}^{m} i(\pi + \theta^j) k_j (z_j - z)$$

and each $c_j = - k_j$.

APPENDIX B
CONVERGENCE OF CVBEM APPROXIMATOR

Let $\omega \in W_\Omega A$, and $z \in \Omega$.

Let $A(z) = P_m(z) + \sum_{j=1}^{m} k_j(z_j - z) \, Ln(z_j - z)$,

where $k_j = \dfrac{1}{2\pi i} \left[\dfrac{(\omega_{j+1} - \omega_j)}{(z_{j+1} - z_j)} - \dfrac{(\omega_j - \omega_{j-1})}{(z_j - z_{j-1})} \right]$

and $P_m(z) = \omega_1 + \left[\dfrac{\omega_1 - \omega_m}{z_1 - z_m} \right] (z - z_1)$.

Because $\omega(z)$ is analytic of $\bar{\Omega}$, ℓ can be chosen small enough such that any three neighboring (in sequence) nodes z_{j-1}, z_j, z_{j+1} lie within the radius of convergence of the Taylor series about z_j. That is,

$$\omega_{j+1} = \omega_j + \omega_j'(z_{j+1} - z_j) + \omega_j''(z_{j+1})^2/2! + \cdots$$

$$\omega_{j-1} = \omega_j - \omega_j'(z_j - z_{j-1}) + \omega_j''(z_j - z_{j-1})^2/2! + \cdots$$

Thus

$$(\omega_{j+1} - \omega_j)/(z_{j+1} - z_j) = \omega_j' + \omega_j''(z_{j+1} - z_j)/2! + \cdots$$

$$(\omega_{j-1} - \omega_j)/(z_j - z_{j-1}) = -\omega_j' + \omega_j''(z_j - z_{j-1})/2! + \cdots$$

and

$$k_j = \dfrac{1}{2\pi i} \left[\omega_j'' \dfrac{(z_{j+1} - z_{j-1})}{2} + r_j \right]$$

where r_j is the residual terms of the Taylor series such that $r_j \to 0$ in order 2 as $\ell \to 0$. Thus

$$\lim_{\substack{m\to\infty \\ \ell\to 0}} A(z) = \lim_{\substack{m\to\infty \\ \ell\to 0}} \left[P_m(z) + \sum_{j=1}^{m} \frac{1}{2\pi i}\, \omega_j'' \, \frac{(z_{j+1} - z_{j-1})}{2}\, (z_j - z)\, Ln\,(z_j - z) \right.$$

$$\left. + \sum_{j=1}^{m} \frac{1}{2\pi i}\, r_j\, (z_j - z)\, Ln\,(z_j - z) \right]$$

Evaluating terms,

$$\lim_{\substack{m\to\infty \\ \ell\to 0}} P_m(z) = \lim_{\ell\to 0}\left[\omega_1 + \left(\frac{\omega_1 - \omega_m}{z_1 - z_m} \right)(z - z_1) \right] = \omega_1 + \left. \frac{d\omega}{d\zeta} \right|_{z_1} (z - z_1)$$

and therefore

$$\lim_{\substack{m\to\infty \\ \ell\to 0}} A(z) = \omega_1 + \left. \frac{d\omega}{d\zeta} \right|_{z_1} (z - z_1) + \frac{1}{2\pi i} \int_{\Gamma} \frac{d^2\omega}{d\zeta^2} (\zeta - z)\, Ln\,(\zeta - z)\, dz$$

Integrating by parts,

$$\int_{\Gamma} \frac{d^2\omega}{d\zeta^2} (\zeta - z)\, Ln\,(\zeta - z)\, dz = (\zeta - z)\, Ln\,(\zeta - z)\, \left. \frac{d\omega}{d\zeta} \right|_{\Gamma}$$

$$- \int_{\Gamma} \frac{d\omega}{d\zeta} (1 + Ln\,(\zeta - z))\, d\zeta$$

where

$$(\zeta - z)\, Ln\,(\zeta - z)\, \left. \frac{d\omega}{d\zeta} \right|_{\Gamma} = 2\pi i\, (z_1 - z)\, \left. \frac{d\omega}{d\zeta} \right|_{z_1}$$

APPENDIX C

THE APPROXIMATE BOUNDARY FOR ERROR ANALYSIS

Best Approximation Method used analytic functions that exactly satisfy the Laplace equation throughout the interior of the problem domain, Ω. As the values of $\hat{\omega}(z)$ approach the values of the exact solution of the boundary value problem $\omega(z)$ for all points z on the problem boundary Γ, then the relative error $|\hat{\omega}(z) - \omega(z)|$ is reduced throughout $\Omega \cup \Gamma$. One approach to reduce relative error is to use $|\hat{\omega}(z) - \omega(z)|$ to locate additional basis functions of the Green's type such that the approximation function is better adapted to fit the boundary conditions of the problem. In this fashion, the approximation function $\hat{\omega}(z)$ converges to the true solution $\omega(z)$ in a fashion analogous to an adaptive integration approach.

Rather than examining relative error values on the boundary, Γ, it is useful to determine an approximative boundary $\hat{\Gamma}$ upon which $\hat{\omega}(z)$ satisfies the given boundary conditions for $\omega(z)$ on $\hat{\Gamma}$. That is, given an approximator $\hat{\omega}(z)$, level curves of constant ϕ or ψ on Γ (where $\omega(z) = \phi + i\psi$ and $\hat{\omega}(z) = \hat{\phi} + i\hat{\psi}$) are compared to level curves of constant $\hat{\phi}$ or $\hat{\psi}$ on $\hat{\Gamma}$ where $\hat{\Gamma}$ is determined by setting the known $\phi = \hat{\phi}$ and $\psi = \hat{\psi}$. The resulting boundary $\hat{\Gamma}$ has the property that $\hat{\omega}(z)$ satisfies the specified boundary conditions on $\hat{\Gamma}$ and $\hat{\omega}(z)$ satisfies the governing Laplace equation in the interior, $\hat{\Omega}$. Consequently, $\hat{\omega}(z)$ is the exact solution to the boundary value problem with the true boundary Γ transformed into the approximative boundary $\hat{\Gamma}$.

The utilization of the approximative boundary provides the following features:

(1) An exact solution of the subject boundary value problem is provided for the transformation of the problem boundary.

(2) The approximative boundary can be visually compared to the true boundary as to closeness of geometric fit.

(3) Basis functions can be added to Γ to determine a more refined approximation $\hat{\omega}(z)$ so that $\hat{\Gamma}$ is geometrically closer to Γ in regions of high discrepancy.

(4) The engineer works with a displacement of the problem boundary rather than examining a more abstract relative error propagation along the boundary.

In the following, several mixed boundary value problems of the Laplace equation are approximated by the Best Approximation Method with analytic function basis of the form $f_i = (z - z_i) \, Ln(z - z_i)$. The problems utilize boundaries $\hat{\Gamma}$ which geometrically coincide with lines of constant ϕ or ψ of $\omega(z)$. After developing a $\hat{\omega}(z)$, the approximative boundary $\hat{\Gamma}$ is determined by plotting the corresponding lines of constant $\hat{\phi} = \phi$ and $\hat{\omega} = \omega$ from $\hat{\omega}(z)$. In the accompanying figures, both Γ and the associated $\hat{\Gamma}$ are plotted together so that a direct comparison is seen. Intuitively, as $||\hat{\Gamma} - \Gamma||$ becomes small then necessarily $|\hat{\omega}(z) - \omega(z)|$ is reduced and $||\hat{\Gamma} - \Gamma|| = 0$ implies $\hat{\omega}(z) = \omega(z)$.

Applicaton C-1.

This problem approximates the classic problem of ideal fluid flow over a cylinder. The exact solution is known to be $\omega(z) = z + 1/z$. The problem boundary is specified to be the upper right quadrant as shown in the figure. The corresponding approximative boundary $\hat{\Gamma}$ is plotted along with the true boundary Γ in Fig. C-1.

Application C-2.

The true usefulness of the approximative boundary concept is illustrated for practical problems where the analytic solution $\omega(z)$ is unknown. Using only the specified boundary conditions, the $\hat{\omega}(z)$ is then determined. In C-2, a large roadway embankment is modeled for heat transfer effects. In this case, a freezing conduit is defined on the right side of the boundary. The approximative boundary $\hat{\Gamma}$ corresponding to a 78-node $\hat{\omega}(z)$ approximation is shown in the figure. A closer look at $\hat{\Gamma}$ and Γ near the conduit is provided in the next application problem.

Application C-3 and C-4.

Similar to Application C-2, these applications examine the matching of $\hat{\Gamma}$ to Γ for two geometrically different conduits. From Figs. C-3 and C-4, $\hat{\Gamma}$ closely matches Γ and $\hat{\omega}(z)$ would therefore be an adequate approximation of $\omega(z)$ for most engineering study purposes.

The applications considered in this section demonstrate the utility of determining an approximative boundary corresponding to solving potential problems using a basis of analytic approximation functions. The approximative boundary $\hat{\Gamma}$ is developed by plotting the level curves of constant potential (or stream function) which match the boundary condition values on the problem boundary Γ.

The error of approximation is manifested by the departure of the approximative boundary from the problem true boundary. Where large spatial discrepancies are observed, additional basis functions are needed to increase the approximation accuracy. The approximative boundary can often be argued to better represent the "as-built" or a more realistic problem boundary then the defined problem boundary. This later idea is especially valid in large scale engineering studies where angle points are generally constructed as grossly rounded edges.

To illustrate the Best Approximation results within the interior of the problem domain (using analytic functions), Figs. C-5 and C-6 show groundwater seepage problems with the approximative boundary, and several streamlines and lines of the constant potential plotted. Because the maximum error magnitude ϵ must occur on the boundary, interior values of $\hat{\omega}(z)$ necessarily differ from $\omega(z)$ in magnitude by less than ϵ.

Although the approximate boundary technique is presented for potential problem solutions using a particular basis of analytic functions, the approach directly extends to other basis and other classes of problems. However, selection of a basis whose elements satisfy the operator is most effective in that the approximate boundary represents departure in matching boundary conditions, whether steady-state or unsteady.

INDEX